T0330112

BEYOND INDIVIDUAL CHOICE

BEYOND INDIVIDUAL CHOICE

TEAMS AND FRAMES IN GAME THEORY

Michael Bacharach

Edited by Natalie Gold and Robert Sugden

PRINCETON UNIVERSITY PRESS   PRINCETON AND OXFORD

Published by Princeton University Press, 41 William Street, Princeton, New Jersey 08540

In the United Kingdom: Princeton University Press, 3 Market Place, Woodstock, Oxfordshire OX20 1SY

Library of Congress Cataloging-in-Publication Data

Bacharach, Michael.

Beyond individual choice : teams and frames in game theory / Michael Bacharach; edited by Natalie Gold and Robert Sugden.
p. cm.
Includes bibliographical references and index.
ISBN-10: 0-691-12005-6 (cloth : alk. paper)
ISBN-13: 978-0-691-12005-8 (cloth : alk. paper)
1. Game theory. 2. Economics, Mathematical. I. Gold, Natalie, 1976- II. Sugden, Robert. III. Title.

HB144.B328 2006
330'.01'5193—dc22 2005049608

British Library Cataloging-in-Publication Data is available

This book has been composed in Palatino

Printed on acid-free paper. ∞

pup.princeton.edu

Printed in the United States of America

10 9 8 7 6 5 4 3 2 1

Contents

Illustrations

Tables

Foreword

IN 2002, until the moment when death swiftly, cleanly and unexpectedly claimed him, while chatting after a shoreline run along his favourite beach in Italy, my husband and lifelong companion Michael Bacharach had been working on a book. (His last work on it had been done on his laptop in a shady spot at the 'bagno' earlier that day.) Michael believed this book bound convincingly together the themes and ideas from many years of enquiry into the foundations of decision theory. He was working on it with relentless passion and urgency. It is tragic that he was not able to finish the project himself, as another three months of life would have allowed. Fortunately, when Michael had begun to write up the chapters, he had started not with the introductory scene-setting ones, but with 'the meat', as he put it to me. Equally fortunately Michael, and those who care about the ideas about which he wrote, have been favoured by the disinterested generosity of an authority in the field, a friend and colleague of Michael's—Bob Sugden. Soon after I returned to England, I heard that Professor Sugden had spontaneously offered to do the work to edit Michael's book to enable publication, if that were possible.

During Michael's lifetime his research for the book was supported by a three-year fellowship from the Economic and Social Research Council, buying out his time from his normal duties. My thanks on his behalf go both to that institution and to his college, Christ Church, and to the Economics Department of Oxford University, for allowing him to be entirely released from his duties to take this fellowship. After Michael's death the ESRC, Christ Church and the Economics Department again provided support to enable the book to be edited and published, and I thank them for that also. Thanks are due above all to the two editors: Robert Sugden, who devoted some of the research time provided to him by his own institution, the University of East Anglia, to the project; and Natalie Gold, who, financed by the ESRC, did much of the routine work of editing, as well as coauthoring the introductory and concluding chapters. Finally, I must mention Richard Baggaley of Princeton University Press who, both during Michael's lifetime and after his death, played a very important positive and creative role in bringing about the conditions that have enabled preparation of a publishable manuscript.

A short biography and some recollections of Michael from some of those who knew him, together with a bibliography, can be found on the following website: http://www.economics.ox.ac.uk/bacharach/index.html. I am very grateful to its editors, Diego Gambetta and Natalie Gold, for having conceived and brought to fruition this happy idea.

The son of friends in Italy, seeking to identify Michael after the dreadful news, queried, 'lui era quello magro ma forte?'—thin but strong. That phrase, an impression from one who knew him only slightly, pinpoints something of Michael's essence, and seems as good a short epitaph as any I could wish for. My own unceasing gratitude is for a lifetime of companionship which I will remember and cherish daily, until death comes for me too.

Elizabeth Fricker
Oxford

Preface

For many years Michael Bacharach was one of the most consistently original and deep-thinking contributors to the theory of games. He died suddenly in the summer of 2002, at the age of sixty-five. For most scholars who die in their sixties, it is only a polite fiction to say that they were still at the height of their powers; but in Bacharach's case, that is the plain truth.

Since the early 1990s he had been developing a theory of how rational individuals solve problems of cooperation and coordination. His approach deviated from the conventional theory of rational choice in two distinct ways. First, it treated the framing of a decision problem—the conceptual scheme within which the decision-making agent 'sees' the problem—as an exogenous variable. Second, it allowed agency to be attributed to groups or 'teams' as well as to individuals. Bacharach had presented some of the main elements of his theory in a scattering of technical papers, written in his characteristically understated style. Other elements were still unpublished. When he died, he was in the middle of writing a book which was to synthesise these ideas and bring them to the attention of a wider audience. This book preserves as much as possible of Bacharach's unfinished work, and reconstructs at least some of the remainder.

The Background to the Book

For most of his professional career, Bacharach's principal research interest was in the foundations of game theory. Game theory studies the decision problems faced by rational individuals who are interacting strategically with one another—in contrast to the traditional approach of neoclassical economics, in which each individual is treated as an independent rational chooser in an environment which he or she takes as given. Bacharach was ahead of his time in seeing the significance of game theory for economics, writing an elegant introduction to the subject in the 1970s (Bacharach 1976). But from the outset he saw that there were deep problems in the theory—problems with the internal consistency of its assumptions, with the indeterminacy of its predictions about behaviour, and with the accuracy of those predictions when they *are* determinate. His research programme was to understand why these problems arise and how they can be resolved.

In the 1980s, finding coherent foundations for game theory was seen as one of the most important theoretical projects in economics. Game theory was understood as a method of finding 'solutions' to formally-defined games. The core solution concept was *Nash equilibrium*—a list of strategies, one for each player, such that it is optimal for each player to follow his listed strategy, given that the other players follow theirs. Game-theoretic analysis worked on the implicit assumption that rational play would lead to a Nash equilibrium, but it was not clear *why* the individual players of a game would or should choose strategies which, in combination, constituted such an equilibrium. It was known that every game (strictly, every finite game) has at least one Nash equilibrium, but many games have more than one. This raised the problem of *equilibrium selection*: if there are multiple equilibria, how do players, reasoning individually, coordinate on the same one? In some games, some combinations of strategies, though making up a Nash equilibrium, seem intuitively implausible—for example, because they rely on players making incredible threats. This observation prompted the *equilibrium refinement* programme of trying to find coherent criteria for excluding implausible equilibria.

Bacharach (1987) made a major contribution to the understanding of the foundations of game theory through a pioneering application of *epistemic logic*—the logic of propositions about knowledge. He explored the implications of what we shall call the *classical* game-theoretic assumptions: that each player of a game maximises expected utility, given the expected behaviour of other players; that players have common knowledge of the game itself; and that they have common knowledge of one another's rationality. Most game theorists had treated it as self-evident that if a unique Nash equilibrium exists, rational players will choose the strategies that lead to it. Bacharach proved that, in general, the classical assumptions do not imply that the players' chosen strategies make up a Nash equilibrium, even if the game has a unique equilibrium. He showed that the argument which leads to that supposedly self-evident conclusion depends on a hidden premise, namely that, for every game, there is a unique 'solution', prescribed by some unspecified form of valid reasoning and thereby accessible to ideally rational players. Bacharach revealed the necessity of this premise and pointed out that no adequate justification had been given for it.

This result was one of a cluster of negative conclusions reached by game theorists in the 1980s. Many theorists were particularly disturbed by a problem that turned out to be central to the equilibrium refinement programme: in a game in which players move sequentially, what it is rational for one player to do may depend on what another player would do in a contingency that would not arise through rational play. The growing realisation that this problem was insoluble within the classical framework led to a loss of confidence in the conception of game theory as a self-contained analysis of ideally rational play. In relation to the practical application of game theory, the problem of equilibrium selection was perhaps more pressing. As game theory began to be put to work in economics, it became increasingly evident that games with multiple equilibria were the rule rather than the exception, and that the classical assumptions did not generate criteria for selecting between equilibria. This problem, too, raised doubts about the coherence of the classical project of identifying ideally rational play in games. If (as had generally been supposed) each game has some particular Nash equilibrium which is its unique rational solution, then in games with multiple equilibria there ought to be some rationally defensible method of selecting one of them as the solution. The failure to find such selection criteria was troubling.

Most game theorists seem to have interpreted these negative conclusions as showing that classical game theory had attributed too much rationality to agents. From the beginning of the 1990s, theorists turned their attention to evolutionary and behavioural models, in which agents are not assumed to be perfectly rational. Instead, it is assumed that they use simple decision-making heuristics that are adapted to the cognitive limitations of human beings, or that their behaviour is the outcome of trial-and-error learning, or that behaviour emerges from blind processes of evolutionary selection. In most of this work, Nash equilibrium has retained its status as the central solution concept of game theory, but it has been reinterpreted. It is treated as a stationary point in a dynamic process in which a game is played recurrently in a population, rather than as the immediate outcome of strategic reasoning by ideally rational agents.

Bacharach remained committed to the older project of investigating rational play in games. In some of his work he relaxed classical rationality assumptions, but he did so by representing cognitive constraints explicitly in models of bounded rationality. Sometimes he built these constraints into his model of reasoning, investigating the nature of valid reasoning, given these constraints. For example, he explored the idea that there are finite limits on the 'depth' to which humans can reason (Bacharach and Stahl 2000). In other work, he built cognitive constraints into the formal representation of decision problems, thus maintaining the principle that each agent's reasoning is valid in relation to the problem she faces. In the introduction we will suggest that his theory of framing has elements of this form of bounded rationality. However, Bacharach seems to have thought that the main flaw in classical game theory was not that it assumed that individuals are better at reasoning than they really are. Rather, its model of valid reasoning was flawed. His principal objective was to reconstruct game theory with a better model of reasoning—not so much a more realistic model of what human beings are capable of, given their limitations, as a truer representation of what valid reasoning is.

We have already described the first component in this reconstruction: Bacharach's epistemic model of rationality and common knowledge, which broke the link between the classical assumptions and Nash equilibrium. The second component was *framing*. According to Bacharach, conventional game theory confuses the world as seen by the theorist with the world as seen by the decision-making agent. In constructing a stylised mathematical model of an interaction, the game theorist imposes a particular conceptual scheme on the world. The analysis of the resulting game implicitly assumes that the agents of the model reason within the scheme that the theorist has imposed. Bacharach argued that it is essential to distinguish between the decision problem that appears in the theorist's model and the problem as the agent represents it to herself. He proposed enlarging the model of a game so that it explicitly includes the agents' *frames*—the sets of descriptions that the players use to represent the problem to themselves. The crucial idea was that an agent can choose, and can think about other agents choosing, only those things for which her own frame provides descriptions. These ideas provided the foundations for Bacharach's *variable frame theory* (Bacharach 1993; Bacharach and Bernasconi 1997), which is described in the introduction.

For Bacharach, one of the merits of variable frame theory was that it helped to explain how people can coordinate their actions without communicating with one another. Ever since Thomas Schelling's *The Strategy of Conflict* (1960), it has been known that human players can solve coordination problems by identifying 'focal points'. (For example, Schelling asked each of a set of respondents to name a place in New York City in which to meet another respondent who had been given the same task; more than half of his sample chose Grand Central Station.) Although the idea of a focal point seems intuitive, game theorists had never succeeded in explaining precisely what a focal point is, or why it is rational for each player to choose her component of the combination of strategies that constitutes one. Bacharach showed that, in variable frame theory, a coordination problem of the 'name a place in New York City' form can be represented as what he called a Hi-Lo game. (How such a representation is possible will be explained in the introduction.) In the simplest form of the Hi-Lo game, two players, Player 1 and Player 2, each choose either *high* or *low*. If both choose *high*, each gets a utility payoff of 2. If both choose *low*, each gets 1. Otherwise, each gets 0. In Bacharach's analysis, the focal point of the coordination game is shown to correspond with the Nash equilibrium in which both players choose *high*.

At first sight, this argument seems to resolve the problem of explaining why it is rational to choose focal points. There may be some mystery about why Schelling's respondents chose Grand Central Station rather than, say, the top of the Empire State Building; but, one might think, there is no mystery about why Player 1 and Player 2 each choose *high* in the Hi-Lo game. In presenting variable frame theory in 1993, Bacharach rested his case at this point. However, as he recognised at the time, this solution does not completely solve the original problem. The difficulty is that although it seems obvious that *high* is the unique rational choice for each player in Hi-Lo, we cannot prove this by using only the classical assumptions of game theory.

Bacharach was not the first person to have the idea that Schelling's coordination problems could be transformed into Hi-Lo games by taking account of framing, although his variable frame theory was the first general analysis of this kind of transformation. The basic idea of transforming coordination problems into Hi-Lo games had been proposed previously by David Gauthier (1975) and Robert Sugden (1991). Sugden's paper contained an additional argument (developed in more detail in Sugden 1993, 1995), which showed that, in a certain special sense, choosing *high* in Hi-Lo *is* uniquely rational. Sugden pointed out that the question 'What is it rational to do in the Hi-Lo game?' can be posed in two different ways. In classical game theory, the question is posed for each player *separately* as 'What should I do?'; the recommendations of the theory are addressed separately to each player, as an individual. But it is also possible for the players *collectively* to pose the question as 'What should we do?'; correspondingly, there can be recommendations that are addressed to the players as a collective. To the question 'What should I do?', the correct answer is 'That depends on what the other player can be expected to do'. But the correct answer to the question 'What should we do?' is 'You should each choose *high*'. In other words, if Player 1 and Player 2 'think as a team'—if they construe themselves as components of a single unit of agency—their decision problem has a unique rational solution: they jointly choose the pair of strategies (*high*, *high*).

The idea that teams can be agents had been proposed before by D. H. Hodgson (1967) and Donald Regan (1980), both of whom discussed the Hi-Lo game and argued for the rationality of *high*. The idea of collective agency (or 'plural subjects') had also been developed by Margaret Gilbert (1989), by Bacharach's then Oxford colleague Susan

Hurley (1989), and by Raimo Tuomela (1995). Drawing on these various sources, Bacharach (1999) developed the first—and so far the only—general theory of *team reasoning* in games.

This theory has what may seem to be a serious limitation. It analyses the reasoning that an individual uses if, taking herself to be a member of a team, she asks, 'What should we do?' It also analyses the reasoning that the individual uses if, *not* taking herself to be a member of a team (or, equivalently, taking herself to be the only member of a one-person team), she asks, 'What should I do?' But it does not explain how the individual comes to ask one of these questions rather than the other. To put this another way, the theory does not explain which group the individual *identifies* with. Bacharach was dissatisfied with this limitation, and tried to extend the theory in a way that would make group identification endogenous. His strategy for doing this was to treat group identification as a form of framing, and hence to bring it within the scope of variable frame theory. Thus if an individual sees a decision problem in the 'we' frame, the question 'What should we do?' comes to mind for her; but if she sees it in the 'I' frame, the question that comes to mind is 'What should I do?' In order to complete this theory, Bacharach needed to make empirical assumptions about how the sense of group identity is activated. With characteristic thoroughness and intellectual curiosity, he studied the research literatures of social psychology, anthropology and biology to discover what was known about group identification, and tried to make his theory consistent with this knowledge.

Bacharach seems to have been the first theorist to realise that the concept of team reasoning might have some application to problems of dynamic choice. In theories of dynamic choice, a standard approach is to treat a *person* (understood as a continuing entity through time) as a succession of *transient agents*, each of which exists in a distinct period of time and is in control of the person's decisions in that period. A central problem in the literature of dynamic choice is to investigate the conditions under which the separate decisions of a person's transient agents are consistent with a single rational plan for the person as a whole. Bacharach's idea was that, in relation to one another, transient agents can face problems of coordination and cooperation, and that these problems might be solved by team reasoning. In this case, the analogue of 'What should we do?' is 'What should I, as a continuing person, do at this moment?'; the analogue of group identity is *personal identity*.

Following these interrelated lines of investigation, Bacharach was gradually constructing a distinctive theory of strategic decision-making that is radically different both from classical game theory and from more recent evolutionary and behavioural theories. His approach makes use of ideas from biology and psychology while, in the spirit of classical game theory, seeking to model processes of valid reasoning. It retains the traditional conception of instrumental rationality, but it dispenses with the equally traditional assumption that rational agency is a property only of individuals.

The Book

The book that Bacharach intended to write would have been a monumental piece of work, bringing together all the themes we have been sketching out. He was working to a plan that broadly followed the history of the development of his ideas. We shall use roman numerals for Bacharach's proposed chapters and arabic numerals for chapters in the book as it now appears.

Chapter I was to describe 'three puzzles of game theory'—the Prisoner's Dilemma, Hi-Lo, and coordination problems of the kind studied by Schelling,—and to discuss general foundational and methodological issues. Chapter II was to present variable frame theory. Chapter III was to show how this theory transforms coordination games into variants of Hi-Lo. Chapter IV was to discuss Hi-Lo, and to show why this game presents a puzzle for classical game theory. Chapter V was to discuss group identification, understood as a particular kind of frame, and to explain how group identification can be primed by certain features of games. Chapter VI was to argue that a human propensity for group identification might be the result of biological natural selection. Chapter VII was to explain the formal theory of team reasoning, with Hi-Lo as the focus of analysis; team reasoning was to be presented as a mode of reasoning that is valid for individuals who identify with a group. Chapter VIII was to look at team reasoning in situations in which the members of a team can communicate with one another. Chapter IX was to present the Prisoner's Dilemma as a game which has the potential to prime either the 'I' frame or the 'we' frame; it would show that team reasoning induces cooperation in this game if the players see themselves as members of a common team, and if they identify with that team.

Bacharach originally planned to write an additional chapter developing the idea that a single person can be understood as a team of transient agents. The argumentative force of this chapter would have been to show that the idea of team reasoning is built into our everyday ideas of personhood. It seems that, at the time of his death, he had decided against trying to incorporate this material into *Beyond Individual Choice*.

In his initial plans, he envisaged his analysis of the person as a reasonably straightforward application of the theory of team reasoning; but it was turning into a much bigger project, involving issues in philosophy and in dynamic choice that were separable from the main themes of the book. Had he lived, the next problem on his research agenda might have been to develop a theory of the person as a team of transient agents.

Bacharach began writing the book at chapter IV. When he died, he had completed drafts of chapters IV, V and VI, and of all but the final sections of chapter VII. Although various references, figures and data were missing, this text was essentially in a finished state—give or take a few rough edges, which he would have smoothed off before publication. For the other chapters, however, he left only successive plans and synopses, and snippets of text which he had written as notes to himself. In trying to piece together what he intended to say, we can also use his published papers and the texts of, or notes for, papers on themes connected with the book that he presented at various conferences. Although the central arguments remain constant across these various plans and presentations, there are successive changes of theoretical detail and variations in emphasis and in argumentative strategy.

Our fundamental aim as editors has been to produce a book that conveys to the widest possible audience as much as can be reconstructed of Bacharach's intended work. We have deliberately chosen not to produce a scholarly compilation of Bacharach's unpublished papers. He was writing this book with the aim of engaging the attention of readers across the social sciences. To be faithful to this intention, we have tried to create a book that can be read by anyone who is interested in the problems of coordination and cooperation or in the foundations of game theory.

As a writer, Bacharach had extremely high standards. All his published work is beautifully crafted. Whether expressed in words or in mathematical symbols, his arguments are always absolutely rigorous; nothing is loose or ambiguous or casual. It would be a disservice to his memory to take material that he regarded only as preliminary notes and to include it in a book that he intended as one of the major works of his professional life. We have therefore included in this book only what he thought of as finished text. To allow the book to be read as a self-contained whole, we have added two sections that make use of Bacharach's notes, but that are written in our own voices. We write as editors who are sympathetic to Bacharach's research project, not as his spokespersons or advocates.

We have not sought to complete the book in the sense of trying to say everything that Bacharach intended to say. To do so would have been a misguided enterprise. That he was unable to complete his book is a loss that neither we nor anyone else can make up. His notes suggest that, at the time he died, he had only a general sense of what was to go in the missing chapters; he was still trying out alternative lines of argument. Those chapters would have taken form only in the process of being written. We cannot pretend to recover them. What we can present is only a fragment of the whole work that Bacharach imagined, edited into a self-contained book.

The relationship between the book as planned by Bacharach and the book as it is now written is summarised in table P.1. The introduction to the present book is written by us as editors. It summarises the material that, as far as we can tell, Bacharach intended to include in his

TABLE P.1
The relationship between Bacharach's plan and book as it now is

Book as planned by Bacharach	Book in its current form
Chapter I: Introduction (not written)	*Preface (by editors)*
Chapter II: Framing (not written)	
Chapter III: The Coordination Problem (not written)	
	Introduction (editors' reconstruction of intended content of chapters I, II and III)
Chapter IV: The Hi-Lo Paradox	Chapter 1: The Hi-Lo Paradox (as written by MB)
Chapter V: Groups	Chapter 2: Groups (as written by MB)
Chapter VI: The Evolution of Group Action	Chapter 3: The Evolution of Group Action (as written by MB)
Chapter VII: Team Thinking	Chapter 4: Team Thinking (as written by MB, but with some sections missing)
Chapter VIII: Organisation (not written)	
Chapter IX: Cooperation (not written)	*Conclusion (editors' reconstruction of intended content of missing sections of chapter VII, chapters VIII and IX, and discussion of MB's ideas on the person)*

chapters I to III. We feel reasonably confident that this introduction is faithful to Bacharach's intentions, since it is based on his previously published papers. However, as we shall explain, there are some points at which variable frame theory, as developed by Bacharach in the early 1990s, does not fully correspond with his later treatment of team reasoning. It is not clear whether he intended to present variable frame theory and team reasoning as two separate but related theories, or whether he was planning to revise the former to make it fully consistent with the latter. Chapters 1, 2, 3 and 4 are, respectively, Bacharach's chapters IV, V, VI and VII. These are as written by Bacharach, except for some copyediting and a few editorial additions. Additions which appear in the main text are marked by square brackets; notes and appendices that we have added are prefaced by asterisks. These chapters are followed by a conclusion in which we try to reconstruct what we can of the intended contents of chapters VIII and IX, and discuss Bacharach's ideas about the person. Here we are more speculative than in the introduction. Bacharach did not write any finished papers on these topics; we are working only from his notes and from our recollections of discussions with him. Our intention is to set out Bacharach's research agenda for readers who may want to pursue the lines of investigation it sets out.

The Editors

We end this preface by explaining how we, the editors, fit into the story of Bacharach's professional life.

From the late 1980s onwards, Robert Sugden and Bacharach were, in a loose sense, research collaborators. Both were working on understanding the foundations of game theory; both were exploring ideas of framing and agency. Although they wrote independently, they exchanged ideas, influenced each other, and generally agreed with each other's conclusions. Differences between their positions were mainly matters of emphasis. In particular, Bacharach set more store on the idea of rationality, wanting to find modes of valid practical reasoning by which problems of coordination and cooperation could (in a genuine sense) be solved. Sugden was less concerned with the validity of team reasoning, treating it only as an idealised model of a form of reasoning which people in fact use, whether justifiably or not; he hoped to integrate team reasoning into a descriptive theory of the evolution of conventions, norms and conceptions of salience.

Natalie Gold had worked with Bacharach since he first taught her as an undergraduate student in 1996; at the time of his death he was supervising her doctoral research. Her investigations of framing and the spread of frames, especially normative frames, are closely related to Bacharach's research programme. While Bacharach relaxed classical rationality assumptions in order to get a truer representation of what valid reasoning is, Gold uses a truer representation of the reasoning process to identify which classical rationality assumptions are violated when agents exhibit framing effects. Her emphasis is more on individual decision-making, but her work on how agents come to be in a particular frame is complementary to Bacharach's theory, in which the mode of reasoning used by players depends on their frame.

For advice and recollections, we are grateful to Ken Binmore, Robin Cubitt, Lizzie Fricker, Gerardo Guerra, Shepley Orr and Daniel Zizzo. We have presented some of the material in the introduction and conclusion at the Rationality and Commitment Workshop in St. Gallen and at the conferences Logic, Games and Philosophy: Foundational Perspectives in Prague and Collective Intentionality IV in Siena. We are grateful to the participants at these meetings for helpful comments.

BEYOND INDIVIDUAL CHOICE

Introduction

THE TITLE THAT Michael Bacharach had planned for his book was *Beyond Individual Choice: An Investigation of Three Puzzles of Game Theory*. In the chapters he wrote before he died, only two of his three puzzles are discussed in detail; the book's subtitle has been changed accordingly. Nevertheless, the structure of Bacharach's argument is linked to three particular games which, he claims, create problems for conventional game theory. In this introduction, which summarises the material that he intended to include in the unwritten chapters I, II and III, we describe these games, explain the senses in which they are puzzling, and lay out in general terms what kind of solution to these puzzles he was looking for. Then we present Bacharach's variable frame theory and show how this takes us at least part of the way to a solution to the first of the three puzzles. We end by comparing Bacharach's approach to this puzzle with those taken by other writers.

1. Three Puzzles of Game Theory

The three types of game that Bacharach presents as puzzles are pure coordination games, 'social dilemma' games of the Prisoner's Dilemma family, and Hi-Lo games. The first two of these types of game are better known than the third, and are more obviously useful as models of real social interactions. We begin with them.

Figure I.1 shows an example of a *pure coordination game* (or, in the terminology that Bacharach sometimes uses, a *Schelling game*). Two players, Player 1 and Player 2, are not allowed to communicate. Each is given the instruction: 'Complete the sentence: A coin was tossed. It came down ——'. If they both complete the sentence in the same way, each will win an amount of money (the same amount irrespective of what they write). If they complete it in different ways, neither wins anything. If we assume that the only possible ways of completing the sentence are 'heads' and 'tails', we get the game shown in figure I.1. If we consider other ways of completing the sentence, such as 'luckily' and 'as I expected', we get a game with more strategies but the same basic structure—that is, a game in which each player chooses one element from the same set of *labels*, in which the pair of payoffs is (1,1) if both choose the same label and (0,0) otherwise. A player's payoffs

		Player 2 heads	Player 2 tails
Player 1	heads	1, 1	0, 0
	tails	0, 0	1, 1

Figure I.1. Heads and Tails

are to be interpreted as measures of what she wants to achieve. (We say more about what payoffs are in section 2.)

The Heads and Tails game we have described has been played, anonymously and for monetary rewards, by ninety British students. Seventy-eight (87 per cent) chose 'heads' (Mehta, Starmer and Sugden 1994). This result matches almost exactly an informal experiment run by Thomas Schelling (1960, pp. 54–58) more than thirty years before, in which forty-two American respondents were asked to name either 'heads' or 'tails' with the objective of giving the same answer as an unknown partner; thirty-six (86 per cent) chose 'heads'. Schelling gave the name *focal point* to the Nash equilibrium on which, in games like this, players coordinate. The puzzle is to explain how people are so successful at playing this type of game. What makes one equilibrium a focal point, and how does this property induce individual players to choose the strategies that make up that equilibrium? On the face of it, there seems no reason to choose one strategy rather than another in Heads and Tails. If each player is just as likely to choose one as the other, there is only a 50 per cent chance of their coordinating. On the evidence of these experiments, the success rate for pairs of human players is just over 75 per cent.[1] How do we humans play this game so well?

This can be interpreted as a straightforwardly practical problem. Coordination problems occur in everyday life. (You and a friend are travelling together and lose each other in a crowded place. Where should you go to meet each other?) Many problems of negotiation and bargaining have a structure that is similar to coordination games, despite involving conflicts of interest. (You and a trading partner are haggling over the price of a car that you want to sell and he wants to buy. Within some band of possible prices, both of you would rather trade than not. Each of you wants to hold out for the best terms that the other will concede, but neither of you wants to block a deal.) In problems like these, it is in both parties' interests to coordinate their strategies. Collectively, we all benefit by our all having whatever general skills and knowledge enable us to solve coordination problems. Individually, given that others have such skills and knowledge, each of us benefits from having them too. So it is a matter of practical interest to find out how coordination problems are solved.

The question 'How do we play this game so well?' can also be interpreted as a problem for game theory, understood as a theory which explains actual human behaviour. If in fact people are successful at coordinating in games like Heads and Tails, we might expect game theory to explain this regularity in behaviour and to predict the particular strategy choices by which coordination is achieved. But conventional game theory cannot do this, because it treats the labels given to strategies as irrelevant to the analysis of a game. The Heads and Tails game described in figure I.1 would be treated as one in which the two strategies are completely symmetrical with each other; whatever information is contained in the strategy labels *heads* and *tails* would be stripped out prior to analysis. Thus the theory cannot arrive at any conclusion which gives *heads* a different status from *tails*. But the fact that has to be explained is that *heads* really does have a different status from *tails*: it is by recognising this difference, whatever it is, that people are able to coordinate on *heads*.

Finally, Heads and Tails poses a problem for game theory, understood as a normative theory of rational choice. Conventional game theory cannot differentiate between *heads* and *tails*, and therefore, in making recommendations about rational action, it cannot tell players to choose *heads* rather than *tails*. Yet it seems that ordinary human players can reason in some way which leads them to choose *heads*. By both reasoning in this way, two players can do better than they can by following the recommendations of conventional game theory. And given the way in which typical human opponents can be expected to reason, *heads* is the better strategy for each player individually. If the conventional theory is based on a correct understanding of rationality, why are its recommendations inferior to those of ordinary human reasoning? It seems that game theory must be using a deficient model of rationality.

The second puzzle is the Prisoner's Dilemma. Figure I.2 shows a typical version of this well-known game. For each player, *defect*

		Player 2	
		cooperate	*defect*
Player 1	*cooperate*	2, 2	0, 3
	defect	3, 0	1, 1

Figure I.2. The Prisoner's Dilemma

strictly dominates *cooperate*. (That is, irrespective of what the opponent chooses, *defect* gives a higher payoff than *cooperate*.) Thus in its explanatory form, conventional game theory predicts that both players will choose *defect*. In its normative form, it recommends *defect* to both players. Yet both would be better off if each chose *cooperate* instead of *defect*.

Is that a puzzle? If, in fact, almost all human players of Prisoner's Dilemma games chose *defect* (and if they construed this choice as rational), it might reasonably be argued that there was nothing to be puzzled about. It would just be an unfortunate fact about rationality that the actions of rational individuals can combine to produce outcomes which, from every individual's point of view, are suboptimal. But the truth is that, in experiments in which people play the Prisoner's Dilemma for money, anonymously and without repetition, the proportion of participants choosing *cooperate* is typically between 40 and 50 per cent.[2] And if one describes the game to ordinary people (or, indeed, to philosophers or to social scientists who have not been trained in economics), one finds a similar division of opinion about what a rational player ought to do. Some people find it completely obvious that the rational choice is *defect*: others are equally convinced that rationality requires each player to choose *cooperate*.

The Prisoner's Dilemma, like Heads and Tails, poses practical problems for us collectively, as citizens. Economic and social life constantly throws up real games of the Prisoner's Dilemma type. (Think of individuals' decisions about whether to vote in elections, whether to contribute to fund-raising appeals for public goods, whether to reduce consumption of carbon fuels, and so on.) It would be better for all of us if each of us was disposed to be cooperative in such games. The evidence shows that some people *do* act on this disposition in some circumstances. If we understood better what factors induced cooperation, we might find ways of structuring the social environment so as to make cooperation more common.

Like Heads and Tails, the Prisoner's Dilemma poses a problem for explanatory game theory. Conventional game theory predicts that players will always choose *defect*, while in fact many players choose *cooperate*: the theory is failing to explain observed behaviour in games. There is a parallel problem for normative game theory. The theory prescribes *defect*, but many people have the strong intuition that *cooperate* is the rational choice. Of course the game theorist may argue that this intuition is mistaken, and insist on the normative validity of the standard analysis. In doing so, the game theorist can point out that any individual player of the Prisoner's Dilemma does better by choosing *defect* than by choosing *cooperate*, irrespective of the behaviour of her opponent. In other words, each individual player can reason to the conclusion: 'The action that gives the best result *for me* is *defect*'. But, against that, it can be said with equal truth that the two players of the game both do better by their both choosing *cooperate* than by their both choosing *defect*. Thus each player can also reason to the conclusion: 'The pair of actions that gives the best result *for us* is not (*defect*, *defect*)'.[3] It seems that normative argument between these two positions leads to a standoff.

In conventional game theory, Heads and Tails and the Prisoner's Dilemma are seen as very different types of game. Heads and Tails has two pure-strategy Nash equilibria, (*heads*, *heads*) and (*tails*, *tails*), and a mixed-strategy equilibrium in which each player chooses each strategy with probability 0.5. The theory cannot explain why one of the pure-strategy equilibria is 'selected' and not the other, but the fact that one of them comes about is not in itself seen as surprising: there is an implicit presumption that rational players will somehow get to Nash equilibrium.[4] If, for whatever reason, the players of Heads and Tails come to expect each other to choose *heads*, that expectation is self-fulfilling. Thus the problem of coordinating behaviour in a game like Heads and Tails can be recast as a problem of coordinating players' mutual expectations. In contrast, either player's choice of *cooperate* in the Prisoner's Dilemma is seen as downright irrational. Even if each player expected the other to choose *cooperate*, each would have an incentive to act contrary to the other's expectation.

Nevertheless, Bacharach claims that, properly understood, the puzzles posed by Heads and Tails and by the Prisoner's Dilemma are closely related. The link between them is the third puzzle: the Hi-Lo game. In general, a Hi-Lo game is a game in which each of two players chooses one element from the same set of labels; if the two players choose different labels, the pair of payoffs is (z, z); for each label i, there is some $a_i > z$ such that the pair of payoffs is (a_i, a_i) if both choose that label; and there is one label j such that a_j is strictly greater than every other a_i. We focus on the particular version of Hi-Lo shown in figure I.3.

		Player 2	
		high	*low*
Player 1	*high*	2, 2	0, 0
	low	0, 0	1, 1

Figure I.3. Hi-Lo

Hi-Lo combines features of Heads and Tails and the Prisoner's Dilemma. Like Heads and Tails, this is a *common-interest game*—that is, a game in which the interests of the players are perfectly aligned, signalled by the fact that, in each cell of the payoff matrix, the two players' payoffs are equal to each other. There are two pure-strategy Nash equilibria, each associated with a different label and coming about if both players choose that label. In this sense, Hi-Lo poses a coordination problem, just as Heads and Tails does: each player wants it to be the case that they both choose the same label. The crucial difference from Heads and Tails is that, in Hi-Lo, one of the equilibria is strictly better than the other for both players. At first sight, this makes the coordination problem in Hi-Lo trivial: it seems obvious that the players should coordinate on the equilibrium they both prefer, namely (*high*, *high*).

Hi-Lo shares with the Prisoner's Dilemma the feature that, of the outcomes that occur if both players choose the same label, one is better than the other for both players. In this sense Hi-Lo poses a cooperation problem: both players benefit by their both choosing *high* rather than *low* just as, in the Prisoner's Dilemma, both players benefit by their both choosing *cooperate* rather than *defect*. The difference is that in the Prisoner's Dilemma, (*cooperate*, *cooperate*) is not a Nash equilbrium while in Hi-Lo, (*high*, *high*) is. It might seem that, because of this difference, the cooperation problem in Hi-Lo is trivial too.

Certainly Hi-Lo does not pose practical problems for ordinary people, either individually or collectively. In experiments in which participants play Hi-Lo games, and in which the *high* and *low* strategies are given neutral labels, the overwhelming majority choose *high*.[5] But Hi-Lo presents a fundamental problem for game theory. From the assumptions that the players are perfectly rational (in the normal sense of maximising expected payoffs) and that they have common knowledge of their rationality, we cannot deduce that each will choose *high*. Or, expressing the same idea in normative terms, there is no sequence of steps of valid reasoning by which perfectly rational players can arrive at the conclusion that they ought to choose *high*. Many people find this claim incredible, but it is true. The argument for its truth will be presented by Bacharach in chapter 1.

If we are prepared to relax the classical assumption of perfect rationality, it is not particularly difficult to construct theories which purport to explain the choice of *high*. After we have stripped out any information contained in their labels, the only difference between the *high* and *low* strategies is that *high* is associated with higher payoffs; because of this, most plausible theories of imperfect rationality predict that *high* is more likely to be chosen than *low*.[6] But it seems unsatisfactory to have to invoke assumptions about imperfections of rationality in order to explain behaviour in such a transparently simple game as Hi-Lo. If we find that standard game-theoretic reasoning cannot tell players how to solve the apparently trivial problem of coordination and cooperation posed by Hi-Lo, we may begin to suspect that something is fundamentally wrong with the whole analysis of coordination and cooperation provided by the standard theory. Conversely, if we could find a form of reasoning which recommends *high* in Hi-Lo, that might provide the key to solving the problems posed by Heads and Tails and by the Prisoner's Dilemma.

2. Reasoning about Games

Crucially, Bacharach's concern is with *models of reasoning*. In the 'Scientific Synopsis' which summarises the arguments he intended to present in the book, Bacharach begins by distinguishing between 'behaviouristic' and 'rational' approaches to the explanation of behaviour.[7] Behaviouristic theories make no reference to reason. In contrast: '[The] rational approach, which is that of classical game theory, sets out to explain behaviour as the outcome of the agent's reasoning or reasons'. Bacharach proposes to follow the rational approach. Here is how he justifies it:

> This research strategy raises many problems . . . and more than once game theorists have felt that it was running into the sands. But it has much to recommend it. People evidently do reason, more or less well. . . . Moreover, some of the reasons that plausibly guide people's behaviour are very general and so have great explanatory power—for example, the reason for choosing an alternative that it maximises expected utility. Finally, reason is a part of nature, so adopting the rational approach does not mean sacrificing the insights of the evolutionary perspective. The method of evolutionary psychology allows us to explain choices by 'reason within nature'.

Reasoning, as understood by Bacharach, is more than a sequence of mental processes. It involves the manipulation of propositions according to well-defined rules. Further, Bacharach is concerned only with modes of reasoning that are *valid*. In propositional logic, a rule of inference—a rule that allows us to derive conclusions from premises—is valid if, whenever the premises are true, so are the conclusions that are derived from them. Here is a simple example of valid reasoning (the propositions above the line are premises, while the proposition below the line is the conclusion):

(1) There are no English mountains over 1,000 metres high.
(2) Snowdon is a mountain which is 1,085 metres high.

Snowdon is not in England.

Bacharach sets out to formulate principles of *practical reasoning*—that is, reasoning which leads to conclusions about what an agent should do—which satisfy analogous criteria of validity. He fleshes out this concept of validity in the following way:

> A rule of inference is held to be valid if it is 'truth-preserving': from any true premises it yields only true conclusions. Similarly, we might call a mode of reasoning in games valid if it is 'success-promoting': given any game of some very broad class, it yields only choices which tend to produce success, as measured by game payoffs.

The fundamental idea is that, in practical reasoning, the agent infers conclusions about what she ought to do from premises which include propositions about what she is seeking to achieve. Such reasoning is *instrumental* in that it takes the standard of success as given; its conclusions are propositions about what the agent should do in order to be as successful as possible according to that standard. If the agent is an individual person, the reasoning is *individually* instrumental. Here is a simple example of individually instrumental reasoning which is valid in Bacharach's sense:

(1) I must choose either *left* or *right*.
(2) If I choose *left*, the outcome will be O_1.
(3) If I choose *right*, the outcome will be O_2.
(4) I want to achieve O_1 more than I want to achieve O_2.

I should choose *left*.

Bacharach interprets the payoffs of a game as specifying what the players want to achieve as individuals (or what counts as success for them as individuals). He assumes that payoffs can be treated as utility indices in the sense of expected utility theory so that, in situations of uncertainty, a player's success is measured by the expected value of her payoff.[8] Thus the following is valid individually instrumental reasoning for Player 1 in Heads and Tails:

> (1) I am Player 1 in Heads and Tails.
> (2) The probability that Player 2 will choose *heads* is 0.6.
> ___
> I should choose *heads*.

The validity of a reasoning schema is independent of the truth or falsity of its premises. In this case there seems to be no rational grounding for premise (2). What the schema shows is that, *if* Player 1 should have a subjective belief that the probability that Player 2 will choose *heads* is 0.6, *then* Player 2 should choose *heads*.

This schema provides an example of what Bacharach calls *best-reply reasoning*: using rules of inference that are valid for individually instrumental reasoning, a conclusion about how one player should act in a game is inferred from premises about the game itself and about how other players will act in it. This is the kind of reasoning that is analysed in classical game theory.

Bacharach, however, proposes to analyse other kinds of valid reasoning: 'My own approach to explaining behaviour in this book is rational; but it is not classical—the reasoning processes are nonstandard'. He accepts that best-reply reasoning is valid for answering questions which take the form 'What should I do, given my standard of success?' But he claims that this is not the only form of valid practical reasoning. He is particularly concerned to investigate modes of reasoning that are instrumental but *not* individual, and which can be used to answer questions of the form 'What should *we* do, given *our* standard of success?' A large part of this book is devoted to this investigation.

Conceptually, the question of whether a mode of practical reasoning is valid is independent of whether human agents act on the conclusions that can be derived by using it. However, it is fundamental to Bacharach's methodology that an analysis of valid reasoning about a game can help to explain the behaviour of real players. He proposes the hypothesis that there is a general tendency for humans to engage in, and to act on, valid practical reasoning.

As the passage quoted in the first paragraph of this section makes clear, Bacharach sees no conflict between this hypothesis and evolutionary explanations of behaviour. He claims that evolutionary selection operates directly on modes of reasoning, and only indirectly on patterns of behaviour. Human beings are endowed with capacities for reasoning that are the product of evolutionary selection; they then apply those capacities to whatever specific decision problems they face. This claim gives us another criterion for assessing modes of reasoning: functionalism. We should expect natural selection to have given us modes of reasoning that were biologically functional in the environment in which they were selected—that is, modes of reasoning which tended to promote the survival and reproduction of the individuals who used them.

It is not *necessarily* true that validity and function go together. When a person engages in practical reasoning, his reasoning operates on propositions that he believes to be true, and it takes his criterion of success as given. By using modes of practical reasoning that are valid, an individual is led to choose actions which promote his conscious objectives, relative to his subjective beliefs. But functionalism is concerned with actual rather than subjectively expected success, and defines success in terms of survival and reproduction, not conscious objectives. In principle, there could be modes of reasoning that are invalid but functional. One possible hypothesis, favoured in the 'heuristics and biases' tradition of cognitive psychology, is that human decision-making is governed by a large number of distinct, context-specific rules of thumb. Such rules of thumb might have been well adapted to the environment in which they evolved, and yet lead modern humans to make invalid inferences in particular contexts.[9] However, the more general and context-independent human reasoning is thought to be, the less credible it is to suggest that there are major divergences between validity and function. Bacharach's hypothesis seems to be that human reasoning operates according to very general rules, and that those rules have been shaped by natural selection to be truth-preserving and success-promoting.

3. Frames

In a lecture in 2001 in which he presented his current ideas about framing, Bacharach began: 'A frame is the set of concepts or predicates an agent uses in thinking about the world. If I see the marks [in figure I.4] as a circle, a triangle and a cross, my frame includes three shape concepts; if as an omicron, a delta and a xi, three letter concepts. I can also see them as both. But not at the same time. One does not just see, but one *sees as*.'[10]

Formally, Bacharach distinguishes between *objects of choice* (for short, *objects*) and *act-descriptions*. Suppose you are taking part in an experiment; you are presented with figure I.4 and asked to choose one of the

Figure I.4. Three objects

three symbols by drawing a circle around it. Objectively, there are three distinct symbols on the page, and you must choose one of them. If we are the designers of the experiment, we might identify these three symbols to ourselves as x_1, x_2 and x_3 (working from left to right), without supposing that, for example, 'x_1' is a term that has ever occurred to you. Then $\{x_1,x_2,x_3\}$ is a set of objects. But, for you, these objects are described by predicates such as *circle, triangle* and *cross*. If you consciously choose to circle the object x_2, you do so by thinking of this action *as* something for which you have a description—say, by thinking of it as 'choosing the triangle'. The *decision problem* that you face can be defined by a set of act-descriptions such as {*choose the circle, choose the triangle, choose the cross*}.

The relationship between objects and descriptions can be represented formally by defining a set S of objects, a set P of predicates suitable for describing objects in S, and a function $E(.)$ which assigns a (possibly empty) subset of S to each predicate in P; if ϕ is a predicate in P, $E(\phi)$ is the *extension* of ϕ in S, that is, the set of objects in S that are described by the predicate ϕ. In the example, we might define $S = \{x_1, x_2, x_3\}$ and your 'frame' (as participant in the experiment) as a set F of shape predicates which includes *circle, triangle* and *cross*.[11] Then $E(circle) = \{x_1\}$, $E(triangle) = \{x_2\}$ and $E(cross) = \{x_3\}$.

Now look at figure I.5. What if you are asked to circle one of the objects in this figure? If your frame contains only shape predicates, and if we denote the objects as y_1, y_2 and y_3, working from left to right, we have $E(cross) = \{y_1,y_2,y_3\}$. Then there is no description under which you can consciously *choose* one of the objects. Bacharach calls this kind of decision problem a *Buridan problem* (after Buridan's ass, which supposedly starved to death while standing equidistant between two piles of hay). However, you can still *pick* one of them arbitrarily (though apparently the ass could not). That is, you can act with the conscious intention of circling *a* cross, without having any particular cross in mind. Bacharach treats 'picking' as an act-description. Thus for a person who has only shape concepts, the decision problem presented in figure I.5 is represented by the singleton {*pick a cross*}. Alternatively, you might conceptualise this problem in terms of the position predicates *left, centre* and *right*, in which case your decision problem is the set of act-descriptions {*choose the left, choose the centre, choose the right*}.

Figure I.5. Three more objects

The same set of objects may be capable of being perceived as two or more distinct decision problems. For example, depending on the frame used by the decision-maker, the decision problem of figure I.4 may be {*choose the circle, choose the triangle, choose the cross*} or {*choose the omicron, choose the delta, choose the xi*}. These two decision problems are alternative ways of describing a single set of objects, namely $\{x_1, x_2, x_3\}$. In describing a given set of objects, a given individual may sometimes use one frame, sometimes another, depending on contextual factors which bring to mind, or *prime*, particular sets of ideas. This feature of real decision-making is hidden in conventional decision theory, in which no distinction is made between objects and act-descriptions.

In the light of Bacharach's analysis of frames, conventional decision theory might be interpreted as making the implicit (but distinctly questionable) assumption that there is a one-to-one relationship between objects and act-descriptions—or, in effect, that the descriptions that a theorist uses in modelling a problem faced by a decision-making agent are the same as those used by the agent himself. On this reading, conventional theory simply assumes away the problem of 'variable frames'.

An alternative reading, suggested by Kenneth Arrow (1982), by Amos Tversky and Daniel Kahneman (1986) and by Frederic Schick (1997), is that the theory rests on an implicit assumption of *extensionality* or *invariance*, to the effect that a given individual's choice from a given set of objects is independent of the descriptions she applies to them. Like other fundamental assumptions of decision theory, invariance might be interpreted as a principle of rationality. Unfortunately for the theory, experimental tests have shown that the behaviour of actual decision-makers often reveals what appear to be systematic violations of rationality principles, and the principle of invariance is no exception. For example, choices over lotteries are influenced by whether outcomes are presented as gains or losses relative to the status quo. Kahneman and Tversky (1979) asked experimental subjects to imagine that 'In addition to whatever you own, you have been given 1,000 [Israeli pounds]', and then to say which they would choose, a certain gain of 500 (A) or a 0.5 chance of gaining 1,000 (B); 84 per cent preferred A to B. A different group of subjects were asked to imagine that they had been given 2,000 and then to say whether they would choose a certain loss of 500 (A′) or a 0.5 chance of losing 1,000 (B′); 69 per cent preferred B′ to A′. If we define objects to be probability distributions over final wealth, the two decision problems are extensionally equivalent, that is, different descriptions of the same set of objects of choice. (If the subject's initial wealth is w, A and A′ both give her $w + 1{,}500$ with certainty, while B and B′ both give her a 0.5 probability of $w + 1{,}000$ and a 0.5 probability of $w + 2{,}000$.) Kahneman and Tversky's interpretation of this phenomenon is that people tend to be risk-averse with respect to changes that are perceived as gains but risk-loving with respect to those that are perceived as losses; because of this, a person's preferences over given final outcomes or *end-states* can be shifted by experimental manipulations which affect her perception of the status quo.

One response is to deny that rational choices must satisfy the principle of extensionality. This is the approach taken by Schick (1997), who argues that standard decision theory overlooks the importance of the agent's *understanding* of the options, a notion which has similarities to Bacharach's descriptions, in activating the values that guide that choice. According to Schick, "We may value an outcome differently under different descriptions of it, but by fixing on this one or that, our understandings determine (select) which of our values count" (p. 55). Hence, because the description may legitimately affect the way an agent values the options, subjects may rationally choose A and B′ in Kahneman and Tversky's experiment (pp. 48–56).

Another response to such evidence is to defend standard rationality assumptions but to deny that end-states are the objects in relation to which invariance should be defined. For example, we might define 'objects of choice' so that the description of such an object specifies, not only the end-state that is achieved, but also the sequence of steps, or 'path', by which that end-state is reached. On this definition, the choice between A and B in Kahneman and Tversky's experiment is *not* extensionally equivalent to that between A′ and B′. That this kind of argument is possible illustrates that invariance can be defined only in relation to a given definition of what counts as an 'object', and any such definition is contestable. Bacharach sums up his discussion of this example by saying:

> So, whether there is a violation of [invariance] depends on how *we*, the theorists, 'cut up the world'. Theorist 1 says we should specify alternatives as end-states, Theorist 2 as paths. Behind this lies: Theorist 2 allows agents to care about paths, Theorist 1 says this is irrational. [The criterion of invariance] can only be applied after resolving a question about what it is rational to care about. For decision theory, there are no unproblematically given 'same things'.

The suggestion here is that a full theory of rational choice would need to address questions about whether particular ways of 'cutting up the world' are rational or irrational, and what it is rational to care about. If these question were left unanswered, the theory would have nothing to say about how an agent's choices ought to relate to the world as it really is (or as it appears to the theorist). However, although

Bacharach occasionally considered such questions,[12] his main concern in this book is with rationality in the specific sense of valid practical reasoning. Validity in reasoning concerns the relationship between premises and conclusions. Since an agent can reason only from premises that are accessible to her, it is inescapable that her premises refer to the world *as it appears to her*. In modelling practical reasoning, we must treat the agent's frame as setting the parameters within which she reasons. As modellers, we can propose hypotheses about how the agent's frame is affected by features of the world as we represent it in our models. This allows us to investigate how the conclusions that the agent reaches by valid reasoning correspond with the world as we have represented it. But we are not entitled to presuppose specific principles of correspondence, such as invariance, on grounds of rationality.

4. Variable Frame Theory and Coordination Problems

The analysis presented in section 3 is one component of Bacharach's *variable frame theory*.[13] In its entirety, this theory is an analysis of rational play in games which takes account of frames. It explicitly models the players' framing of their decision problems and assumes that, from the viewpoint of each player, the framework within which she reasons is given to her by her own frame.

The theory begins with a given game, defined in an entirely conventional way in terms of players, strategies and payoffs. The game is to be understood as representing an interaction between the players as it really is, or as the theorist describes it. We may call this the *objective game*; its *objective strategies* correspond with the 'objects of choice' analysed in section 3. The crucial difference from conventional game theory is that each player's decision problem is *not* the set of objective strategies from which, objectively, she chooses. Rather, it is a set of act-descriptions. Variable frame theory transforms the objective game into a *framed game* (or game in *option form*) in which players choose among act-descriptions. To keep the exposition simple, we will deal only with games with two players, although the theory applies to games with any number of players.

The theory imposes some structure on the predicates that can be applied to objective strategies, and which are used in act-descriptions. On our reading, these assumptions are not to be treated as fundamental components of variable frame theory, or as definitive hypotheses about how people think about games. Rather, they are simplifying assumptions which are intended to capture certain aspects of framing that are particularly significant in relation to the coordination games that Bacharach is analysing.

As in the analysis in section 3, act-descriptions are constructed from *predicates*. The fundamental structural assumption is that these predicates belong to disjoint sets, called *families*. Each family is interpreted as a set of mutually exclusive but not necessarily exhaustive specifications of a potential attribute of objective strategies. Thus each objective strategy is described by no more than one (but perhaps by no) predicate from each family. The extension of a predicate is defined as in section 3.

For example, in a coordination game in which each player is required to select one object from the set of objects in figure I.4, each being rewarded if they both select the same one, the families might correspond to different potential attributes of these objects. Thus we might have the shape family F_1 = {*circle, triangle, square, cross, . . .* }, the Greek letter family F_2 = {*omicron, delta, xi, alpha, . . .* }, the position family F_3 = {*left, right, centre, top, . . .* }, and so on. It is convenient to define the *generic* family F_0 = {*thing*}, interpreted so that the extension of *thing* includes every object in the relevant decision problem. This gives players a general-purpose predicate with which to describe things among which they choose. Throughout the remainder of this section, we will use F_0 in this way.

A *frame* is a set of such families. Each player has a frame. For example, a player who recognises only shape, position and 'thing' has the frame $\{F_0, F_1, F_3\}$. The theory takes some *universal frame* $F = \{F_0, \ldots F_n\}$ as given. Every relevant predicate belongs to one and only one of the families in the universal frame. However, individual players are not necessarily aware of all these predicates. For each family F_i there is a probability $v(F_i)$, the *availability* of that family. For each player, for each family F_i, the probability that that family is in his frame—that it *comes to mind* for him—is $v(F_i)$. These probabilities are independent across families and across players. Notice that these assumptions imply that if one particular predicate comes to mind for some player, then so too do all the other predicates in the same family. It is assumed that $v(F_0) = 1$, guaranteeing that each player's decision problem contains at least one act-description.

Each player's decision problem is constructed from his set of objective strategies, the predicates which in fact apply to those strategies, and his frame. Act-descriptions take one of two forms: *choose the q*, where q is a predicate whose extension in the set of objective strategies for the relevant player is a singleton, and *pick an r*, where r is a predicate whose extension in this set contains two or more strategies. If a player acts on a description of the latter kind, it is assumed that each of the eligible objective strategies has the same probability of being selected. By specifying each player's decision problem in this way, Bacharach is making an assumption that (as is acknowledged in Bacharach and Bernasconi 1997, p. 7) is rather restrictive: predicates from distinct families cannot be combined in act-descriptions. For example, if *red* is a colour predicate and *square* is a shape predicate, the model does not allow *choose the red square* as an act-description.

Each player is assumed to have beliefs about his opponent's frame. Those beliefs are given by the availabilities of the families that have come to mind for the first player. For example, suppose $F = \{F_0, F_1, F_2, F_3\}$, where F_1, F_2 and F_3 are the shape, Greek letter and position families. Suppose the availabilities of these families are $v(F_0) = 1$, $v(F_1) = 0.8$, $v(F_2) = 0.3$ and $v(F_3) = 0.6$. Suppose that, for some player, only the generic, shape and position families have come to mind. Then that player assigns probability 1 to the event that the generic family has come to mind for his opponent, 0.8 to the event that the shape family has come to mind for her, and 0.6 to the event that the position family has come to mind for her. But, not having the Greek letter family in his own frame, he cannot conceive of that family coming to mind for his opponent: he reasons as if that family of predicates did not exist.

In our formal presentation of the theory, we restrict ourselves to the case in which every family in F has availability 1 for each player. We will give only an intuitive account of how the theory can be generalised; for more detail, the reader can consult Bacharach (1993) and Bacharach and Bernasconi (1997).

The logic of the theory is perhaps best communicated through examples. We follow Bacharach in taking most of our examples from a particular class of pure coordination games, which he calls *blockmarking* games. A blockmarking game has two players, Player 1 and Player 2. Each in turn is shown the same tray of wooden blocks, of the kind that children in old-fashioned families play with. Each player, out of sight of the other, marks one of the blocks in a way that is invisible to the other player until the end of the game. Each player's payoff is 1 if both players mark the same block, and 0 otherwise.

Our first example is the Four Cubes blockmarking game. There are four blocks on the tray: three red cubes and one blue cube. The red cubes are *nondescript*. 'Nondescript' is Bacharach's technical term for objects which, though clearly distinct and perhaps not absolutely alike, do not differ in ways that can be described using predicates that can come to mind for the players. Using x_1, x_2 and x_3 respectively to denote the objective strategies of choosing each of the three red cubes, understood as distinct objects, and x_4 to denote the strategy of choosing the blue cube, the objective game of Four Cubes is as shown in figure I.6. If we strip out any information contained in the labels, the four strategies are completely symmetrical with one another. Intuitively, it seems obvious that each player ought to mark the blue cube, but this conclusion cannot be derived within any theory that treats figure I.6 as an adequate representation of the game.

But now consider the framed game. Suppose that $F = \{F_0, F_1\}$ where F_1 is the colour family {*red, blue, yellow, green, . . .* }. Assume that both families have availability 1. Then, for each player, there are three act-descriptions: *pick a thing, pick a red*, and *choose the blue*. The resulting framed game is shown in figure I.7. The entries in the cells are expected payoffs. (For example, if either player chooses *pick a thing*, the probability that the players select the same object is 0.25, and so the expected payoff is 0.25 too.) Notice that the framed game, like the objective game, is a common-interest game: this follows immediately from the fact that the payoff to each player is simply the probability that the two players select the same object. Notice also that, in both games, there is a pure-strategy Nash equilbrium if and only if both players choose the same objective strategy (in the objective game) or the same act-description (in the framed game). However, there is a crucial difference

		Player 2			
		x_1	x_2	x_3	x_4
Player 1	x_1	1, 1	0, 0	0, 0	0, 0
	x_2	0, 0	1, 1	0, 0	0, 0
	x_3	0, 0	0, 0	1, 1	0, 0
	x_4	0, 0	0, 0	0, 0	1, 1

Figure I.6. Four Cubes as an objective game

		Player 2		
		pick a thing	*pick a red*	*choose the blue*
Player 1	*pick a thing*	0.25, 0.25	0.25, 0.25	0.25, 0.25
	pick a red	0.25, 0.25	0.33, 0.33	0, 0
	choose the blue	0.25, 0.25	0, 0	1, 1

Figure I.7. Four Cubes as a framed game

between the games. In the objective game, the four pure-strategy equilibria give the same payoffs as one another. In the framed game, one of the three pure-strategy equilibria—the one in which each player opts for *choose the blue*—gives higher payoffs than the others (and, in fact, higher payoffs than any other combination of act-descriptions). The framed game is a variant of Hi-Lo, with *choose the blue* corresponding to *high*. If it can be shown that *high* is the unique rational choice in Hi-Lo, Bacharach's analysis explains the rationality of marking the blue cube in Four Cubes.

Formally, variable frame theory picks out the pure-strategy Nash equilibria of the framed game; these are called *variable frame equilibria*.[14] In Four Cubes, there are three variable frame equilibria: (*pick a thing, pick a thing*), (*pick a red, pick a red*), and (*choose the blue, choose the blue*). Bacharach then defines a *solution* as a subset of the set of variable frame equilibria, arrived at by using two principles of equilibrium selection. The first principle is that of 'symmetry disqualification'. As this principle is not relevant for Four Cubes, we postpone discussing it. The second principle, *payoff dominance*, stipulates that an equilibrium is not a solution if some other equilibrium is Pareto-superior to it. (One profile p is *Pareto-superior* to another profile q if, for every player, the payoff under p is at least as high as the payoff under q, and if there is at least one player for whom the payoff under p is strictly greater. A profile p is *Pareto-optimal* in some set S if there is no profile q in S such that q is Pareto-superior to p.) Applying this principle, the solution of Four Cubes contains only one profile, (*choose the blue, choose the blue*).

In his early papers on variable frame theory, Bacharach does not claim to explain why rational players, acting as individual agents, follow the dictates of the principle of payoff dominance. Instead he hedges his bets, saying only that '[w]hether or not [payoff dominance] describes the outcomes of interactions between ideal spirits, it is strongly commended by common sense' (1993, note 4). Our interpretation of passages like these is that Bacharach recognised that payoff dominance could not be justified by appeal to classical game-theoretic assumptions of rationality and common knowledge. He was looking for an alternative justification of this principle, but his thinking on this issue was still in a state of flux; he preferred not to commit himself to any particular justification. He later came to see his theory of team reasoning as explaining why payoff-dominant equilibria are selected in games like Four Cubes. However, as will emerge later in the book, this theory does not imply that payoff-dominant equilibria are selected in *all* games. (Roughly speaking, the idea is that the players of a game reason according to payoff dominance if each of them 'identifies' with the group which comprises those players. Some games are more likely than others to prompt the relevant sense of group identification.) It seems, then, that Bacharach came to see the payoff dominance principle as applying only to a subclass of framed games. Since we do not know whether (and if so, how) he intended to revise variable frame theory to reflect his ideas about group identification, we can only present the theory in its original form.

Our second example, Three Cubes and a Pyramid, allows us to explain the principle of symmetry disqualification. This is another block-marking game with four blocks. There is a red cube, a blue cube, a yellow cube and a green pyramid. The objective game has exactly the same payoff matrix as Four Cubes. Intuition points to the green pyramid as the focal point for coordination, but nothing in the objective game distinguishes this strategy from the others. But suppose that $F = \{F_0, F_1, F_2\}$ where F_1 is the colour family {*red, blue, yellow, green, . . .* } and F_2 is the shape family {*cube, pyramid, sphere, . . .* }. Assume that all three families have availability 1. Reasoning as before, there are seven act descriptions for each player: *pick a thing, pick a cube, choose the pyramid, choose the red, choose the blue, choose the yellow* and *choose the green*. Correspondingly, there are seven variable frame equilibria. The principle of payoff dominance allows us to eliminate the two 'pick a . . .' equilibria, but that still leaves us with five equilibria, each giving the same pair of payoffs (1,1).

Bacharach deals with problems of this general kind by defining the following concept of *isomorphism* between act-descriptions. Suppose that p and q are distinct predicates from the same family and that, for a given player, the number of objective strategies with the property p is equal to the number with the property q. Then the act-descriptions *pick a p* and *pick a q* are isomorphic. Similarly (the case in which exactly one strategy has each property) *choose the p* and *choose the q* are isomorphic. Bacharach stipulates, as a principle of equilibrium selection, that in a solution of a pure coordination game, a player may not choose any act-description that is isomorphic with another. This is the principle of *insufficient reason* or *symmetry disqualification*. He justifies this principle by arguing as follows. Suppose that, for some player, *pick a p* and *pick a q* (or *choose the p* and *choose the q*) are isomorphic. Then, for any valid argument which recommends one of these act-descriptions, there must be an equally valid analogous argument which recommends the other. But (since predicates from the same family are mutually exclusive) the conclusions of any such pair of arguments would be mutually contradictory.

Returning to Three Cubes and a Pyramid, the act-descriptions *choose the red, choose the blue, choose the yellow* and *choose the green* are debarred by the principle of symmetry disqualification. After excluding these, we arrive at the framed game shown in figure I.8. One of the three surviving equilibria, (*choose the pyramid, choose the pyramid*), is Pareto-superior to the others, and hence is the unique solution.

By means of a third example, we introduce some of the additional issues that arise when not all families of predicates have availability 1. Large and Small Cubes is a blockmarking game with ten blocks: seven red cubes and three blue ones. All but one of the cubes are exactly equal in size, but one of the red cubes is slightly larger than the others. This difference in size is such that a typical player might notice it, but might not. The objective game is a pure coordination game with ten strategies for each player. The framed game is more complicated than in our previous examples.

Suppose that $F = \{F_0, F_1, F_2\}$, where F_1 is the colour family {*red, blue, yellow, green, . . .* } and F_2 is the size family {*large, small, . . .*}. Suppose that F_0 and F_1 have availability 1, but F_2 has availability p, where $0 < p < 1$. First, consider the framed game as perceived by a *non-noticer*—a player for whom the size family does not come to mind. For the non-noticer, there are three act-descriptions: *pick a thing, pick a red* and *pick a blue*. He feels certain that these act-descriptions are recognised by his opponent, and cannot conceive of the opponent recognising any others. Thus, the non-noticer's framed game is as shown in figure I.9.

		Player 2		
		pick a thing	*pick a cube*	*choose the pyramid*
	pick a thing	0.25, 0.25	0.25, 0.25	0.25, 0.25
Player 1	*pick a cube*	0.25, 0.25	0.33, 0.33	0, 0
	choose the pyramid	0.25, 0.25	0, 0	1, 1

Figure I.8. Three Cubes and a Pyramid as a framed game

		Player 2		
		pick a thing	*pick a red*	*pick a blue*
	pick a thing	0.1, 0.1	0.1, 0.1	0.1, 0.1
Player 1	*pick a red*	0.1, 0.1	0.14, 0.14	0, 0
	pick a blue	0.1, 0.1	0, 0	0.33, 0.33

Figure I.9. Large and Small Cubes as perceived by a non-noticer

This is another variant of Hi-Lo. Considered as a game in its own right (which is how it appears to the non-noticer), this game has three variable frame equilibria: (*pick a thing, pick a thing*), (*pick a red, pick a red*) and (*pick a blue, pick a blue*). Since (*pick a blue, pick a blue*) is Pareto-superior to the other two equilibria, this is the unique solution.

Now consider how the game appears to a *noticer*—a player for whom the size family *does* come to mind. For the noticer, there are five act-descriptions: *pick a thing, pick a red, pick a blue, pick a small* and *choose the large*. She is certain that the first three of these act-descriptions are recognised by her opponent, but she is uncertain whether he recognises the act-descriptions which refer to *large* and *small*. Bacharach assumes that, in deciding how to act, the noticer treats the behaviour of a non-noticing opponent as given. It is a necessary condition for a variable frame equilibrium in the game as a whole that the noticer's chosen act-description is optimal for her, given that the act-descriptions chosen by non-noticers constitute a variable frame equilibrium in the game as it appears to them. A corresponding condition applies to solutions. So in the solution of the game as a whole, noticers act on the belief that non-noticers opt for *pick a blue*.

Given this belief, what should a noticer do? Given the criterion of Pareto optimality, it is clear that the noticer should opt either for *pick a blue*, which coordinates with the non-noticers, or for *choose the large*, which would be the unique solution in a game in which everyone is a noticer. If the value of p is sufficiently low, the noticer should be guided by what non-noticers can be expected to do, and so should opt for *pick a blue*; if this value is sufficiently high, she should opt for *choose the large*. But what is the critical value of p?

First, consider the case in which both noticers and non-noticers opt for *pick a blue*. Then the expected payoff to each player is $1/3$, irrespective of whether she is a noticer or a non-noticer. Now consider the alternative case in which noticers opt for *choose the large* while non-noticers opt for *pick a blue*. Then the expected payoff to each player is 1 in a game between two noticers, $1/3$ in a game between two non-noticers, and 0 in a game between a noticer and a non-noticer. Since the probability of playing against a noticer is p, the expected payoff to a noticer is p, while the expected payoff to a non-noticer is $(1 - p)/3$. The critical value at which the expected payoff *to noticers* is the same in the two cases is $p = 1/3$.

However, we might instead look for the value at which the expected payoff *to players in general* is the same in the two cases. In the first case, in which everyone opts for *pick a blue*, the expected payoff to every player is $1/3$. In the second case, since the probability that a player is a noticer is p, the expected payoff to a player for whom it is not known whether or not she is a noticer is $p^2 + (1 - p)^2/3$; the critical value at which this is equal to 1/3 is $p = 1/2$.

Which of these two analyses should we use? The analysis which leads to the critical value $p = 1/3$ asks which combination of act-descriptions for noticers maximises players' expected payoffs, *conditional on their being noticers*. The analysis which leads to the critical value $p = 1/2$ asks which such combination maximises *unconditional* expected payoffs. In his presentations of variable frame theory, Bacharach favours the first analysis. He does so, however, in the context of a theory in which payoff dominance is treated as a principle of equilibrium selection. It is not clear whether he intended to retain this analysis when integrating variable frame theory with his later theory of team reasoning. In the perspective of a theory of team reasoning, it seems more natural to look for a plan which is optimal for the team as a whole, where the 'team' includes both noticers and non-noticers. This approach implies the maximisation of unconditional expected payoffs, and hence that the critical value is $p = 1/2$.[15]

Our final example is Heads and Tails itself. As far as we know, Bacharach never analysed this particular game, but he did apply variable frame theory to a game with exactly the same formal structure; we are merely transferring his analysis from the one game to the other.[16] The objective game of Heads and Tails is a pure coordination game with two objective strategies for each player. Let us denote these strategies x_1 (the one that is generally called 'heads') and x_2 ('tails'). What assumptions can credibly be made about players' frames? As before, we define the generic frame $F_0 = \{thing\}$, and assume that this has availability 1. We must surely also assume that the family of coin faces, $F_1 = \{heads, tails\}$, has availability 1. Of course, these assumptions provide no help in solving the coordination problem, since the principle of insufficient reason would debar would-be solutions in which players opt for *choose the heads* or *choose the tails*. If the universal frame is $\{F_0, F_1\}$, the unique solution is (*pick a thing, pick a thing*), and so the probability that the players coordinate is 0.5.

Most people's intuitions seem to be in accord with Schelling's (1960, p. 64) suggestion that 'heads' differs from 'tails' in that it has 'some kind of conventional priority': 'heads' somehow seems more important than, or to come before, 'tails'. How can this be represented in variable frame theory? Bacharach's approach is to define a singleton family $F_2 = \{prominent\}$. For simplicity, we assume that this has availability 1. Then if $F = \{F_0, F_1, F_2\}$, each player has two act-descriptions that are not debarred by the principle of insufficient reason: *pick a thing* and *choose the prominent*. The framed game shown in figure I.10 is the result.

		Player 2	
		pick a thing	*choose the prominent*
Player 1	*pick a thing*	0.5, 0.5	0.5, 0.5
	choose the prominent	0.5, 0.5	1, 1

Figure I.10. Heads and Tails as a framed game

This is yet another variant of Hi-Lo, in which *choose the prominent* corresponds to *high*.

Notice that this resolution of the coordination problem of Heads and Tails depends critically on the assumption that there is a family of predicates (specifically, F_2) such that one objective strategy is described by a predicate from this family (x_1 is described by *prominent*), while the other objective strategy is not described by *any* predicate from that family. In particular, the resolution would break down if F_2 also contained the predicate *not prominent* (since then the principle of insufficient reason would debar the act-description *choose the prominent*). So Bacharach is committed to the hypothesis that *prominent* (or some analogous predicate) can come to mind for a player while *not prominent* does not. This hypothesis might be supported by the argument that, in terms of human cognition, predicates which take the form *not P* are less easily processed than ones that take the form *P*, and hence less likely to come to mind.[17]

In the preface, we suggested that variable frame theory contained elements of bounded rationality. Its analysis of Heads and Tails illustrates what we have in mind. As a matter of empirical psychology, it may be true that a person can have mental access to some predicate *P* but not to *not P*. Still, one might hesitate to call a person 'rational' if he deliberates about *P* without being aware of *not P*. Bacharach himself admits to 'ambivalence' about whether his theory assumes limitations of rationality. His aspiration is '[to show] coordination in favourable cases to be the product of an involuntary processing stage which is neither rational nor irrational, and good game-theoretic reasoning' (1991, p. 42). In other words, he wants to draw a distinction between the frame in which a decision problem is viewed and the reasoning that is applied to that problem. He wants to be able to say that, for the agents in his theory, their frames are supplied by involuntary psychological processes, to which the concept of rationality does not apply, while their reasoning is valid. However, as the case of Heads and Tails seems to show, this distinction is not always easy to maintain.

5. Other Theories of Coordination

Bacharach's analysis of coordination is offered as a resolution of a puzzle first posed by Schelling in *The Strategy of Conflict* (1960). Schelling's book marks the first recognition of the significance of pure coordination games. In it, he shows that (what was then) conventional game theory cannot justify any solution to these games other than each player's choosing among the strategies at random; but he also shows that human players are often remarkably successful at coordinating on particular focal points. He tries to explain what characteristics make these particular pairs of strategies stand out, and how this form of 'standing out' (which he calls *prominence* and which is now usually called *salience*) induces players to choose those strategies.

Generations of game theorists have been influenced by Schelling's work. The concepts of 'focal point' and 'salience' have become part of the tool-kit of game theory. Everyone who reads Schelling's book comes away with an intuitive sense of what makes an equilibrium into a focal point, and that intuition has become part of game theory's folk wisdom. Despite that, focal points have never been integrated into the formal structure of the theory. Although theorists often invoke notions of salience when dealing with otherwise intractable problems of equilibrium selection, they do so in a spirit of *faute de mieux*. The consensus among game theorists seems to be that Schelling has discovered a genuine regularity in the behaviour of human game players, but that neither he nor anyone else has provided an adequate theoretical analysis of what that regularity is, still less an adequate explanation of why it occurs. Bacharach sees variable frame theory as filling that gap: in the final version of his plan for chapter I, he claims that his theory 'solves a long-standing puzzle about coordination by demystifying the notion of focal points and showing formally how focussing works'.

In fairness to Schelling, however, we should recognise that he offers his own analysis of focal points. Although this analysis is not presented in the language of mathematics or formal logic, it is not wholly mysterious. The difficulty for the formally-trained reader is that Schelling's discussion relies heavily on metaphors. Two metaphors are used particularly often. One is the *meeting of minds*, with the suggestion that the process by which players solve coordination problems involves the simultaneous reaching out of two minds towards one another, each trying to find some way of thinking about the problem that can be aligned with the other's way of thinking about it. The other is the *riddle*, with the associated idea of the *clue*. The idea here is that each player reasons about a coordination problem as if it were a riddle with a definite solution, put there deliberately to be recognised by each of them; each looks for this solution, following whatever clues he or she can find.

The idea of the riddle links with one of the guiding ideas of classical game theory, that every well-defined game has a determinate 'solution', discoverable by sufficiently rational players. Although Schelling rejects the traditional game-theoretic model of reasoning as too restrictive, he maintains the assumption of the existence of determinate solutions, at least for coordination games. For Schelling, this assumption is not justified by theoretical considerations about what can and cannot be deduced by valid reasoning. Rather, it is justified by the empirical fact that human players *can* coordinate their strategies in these games. A rational player, recognising this fact, should set about looking for whatever clues his opponent (or, as Schelling prefers to say, 'partner') is likely to use:

> Thus a normative theory of games . . . has a component that is inherently empirical; it depends on how people can coordinate their expectations. It depends therefore on skill and on context. The rational player must address himself to the empirical question of how, in the particular context of his own game, two rational players might achieve tacit coordination of choices, if he is to find in the game a basis for sharing an *a priori* expectation of the outcome with his partner. (1960, p. 285)

In Schelling's accounts of how players solve coordination games, each player begins by assuming that the game he is playing has a determinate solution in which he successfully coordinates with his partner. Each player then tries to find that solution. In carrying out this search, a sensible player will range as widely as he can over the various criteria that might be used to identify a solution, and be as catholic as he can in his conception of what counts as reasoning, allowing room for 'imagination' as well as 'logic' (p. 57). At the same time (and just as in the case of solving a riddle), putative solutions which come to mind more readily will be favoured. For example, consider the pure coordination game in which each player's instruction is 'Name a positive number' and each wins a prize if they both name the same one. According to Schelling:

> Experiments . . . demonstrate that most people, asked just to pick a number, will pick numbers like 3, 7, 13, 100, and 1. But when asked to pick the same number the others will pick when the others are equally interested in picking the same number, the motivation is different. The predominant choice is the number 1. And there seems to be good logic in this: there is no unique 'favored number'; the

		Player 2	
		I	*II*
Player 1	*i*	10, 10	0, 5
	ii	5, 0	10, 10

Figure I.11. A game discussed by Schelling

> variety of candidates like 3, 7, and so forth, is embarrassingly large, and there is no good way of picking the 'most favorite' or most conspicuous. If one then asks what number, among all positive numbers, is most clearly unique, or *what rule of selection would lead to unambiguous results*, one may be struck with the fact that the universe of all positive numbers has a 'first' or 'smallest' number. (1960, p. 94; emphasis in original)

Here is another example. Discussing the game shown in figure I.11, Schelling suggests that a perceptive player would reason as follows:

> Comparing just (*i*, *I*) and (*ii*, *II*) my partner and I have no way of concerting our choices. There must be some way, however, so let's look for it. The only other place to look is in the cells (*ii*, *I*) and (*i*, *II*). Do they give us the hint we need to concert on 10 apiece? Yes, they do; they seem to "point toward" (*ii*, *II*). They provide either a reason or an excuse for believing or pretending that (*ii*, *II*) is better than (*i*, *I*); since we need an excuse, if not a reason, for pretending, if not believing, that one of the equilibrium pairs is better, or more distinguished, or more prominent, or more eligible, than the other, and since I find no competing rule or instruction to follow or clue to pursue, we may as well agree to use this rule to reach a meeting of minds. (1960, pp. 297–98)

As the second passage makes particularly clear, Schelling imposes no standards of logical validity on the reasoning which recommends a focal point to the players of a game. Indeed, one might question whether this is *reasoning* at all, if excuses are as admissible as reasons, pretences as admissible as beliefs. What Schelling has in mind are perhaps better thought of as associations of ideas than as chains of reasoning. One of his favourite rhetorical devices is to defend an apparently arbitrary focal point by asking a question of the form 'If not here, where?' or 'What other clue is there?' (1960, pp. 66, 70, 112, 300). The idea is that a rational player uses the best clues he can find, however good or bad these may be in absolute terms: 'beggars cannot be choosers when fortune gives the signals' (p. 300).

By now it should be clear that Schelling's approach to the analysis of coordination games is very different from Bacharach's. Bacharach's aim is to find formal representations of modes of valid reasoning that are accessible to game-players. As we explained in the preface, Bacharach was one of the first theorists to grasp the full implications of the fact that, if 'rationality' is understood as game theory has traditionally understood it, there is no guarantee that games have determinate solutions. It is fundamental to his approach that the existence of determinate solutions should *not* be treated as axiomatic. Instead we should model valid reasoning as explicitly as possible, leaving open the question of whether it will or will not identify unique solutions for particular games. Seen in this perspective, Schelling's assumption that coordination games have determinate solutions seems question-begging, and his lack of concern about the logical validity of players' reasoning seems (at the very least) unrigorous. Still, it may be misleading to think of variable frame theory as *demystifying* Schelling's account. Bacharach's and Schelling's theories are perhaps better thought of as radically different explanations of a common phenomenon.

Bacharach and Schelling are both game theorists who are challenging some of the conventional assumptions of the theory they have inherited. But, to a greater degree than Schelling,[18] Bacharach is trying to assimilate focal points to game theory: he is trying to extend classical game theory in a way that retains what he sees as its essential features while explaining how people solve coordination problems. The writers whose work we discuss in the rest of this section can all be interpreted as pursuing this assimilationist project.

As far as we know, the first attempt to explain focal points in terms of something close to standard game-theoretic reasoning was by David Lewis (1969, pp. 24–36). Lewis summarises Schelling's experimental results as follows:

> It turns out that sophisticated subjects in an experimental setting can often do very well—much better than chance—at solving novel coordination problems without communicating. They try for a coordination equilibrium that is somehow *salient*: one that stands out from the rest by its uniqueness in some conspicuous respect. (p. 35)

He then offers a theoretical explanation of these observations:

> How can we explain coordination by salience? The subjects might all tend to pick the salient as a last resort, when they have no stronger ground for choice. Or they might expect each other to have that tendency, and act accordingly; or they might expect each other to expect each other to have that tendency and act accordingly, and

> act accordingly; and so on. Or—more likely—there might be a mixture of these. Their first- and higher-order expectations of a tendency to pick the salient as a last resort would be a system of concordant expectations capable of producing coordination at the salient equilibrium. (pp. 35–56)

This account combines two conceptually distinct concepts of salience. The first concept is built into Lewis's official definition of 'salience': that of standing out from the rest by being unique in some conspicuous respect—for short, *conspicuous uniqueness*. The second concept, which has been called *primary salience*, is the property of being picked as a last resort. An option has primary salience in a decision problem if, in the absence of any other reasons for choosing one option rather than another, decision-makers tend to pick it. As derivatives of this concept, we can define *secondary salience* as the property of appearing to decision-makers as having primary salience for other decision-makers, and so on.[19] Lewis assumes that the determinants of primary salience are generally known, so that primary salience and secondary salience go together, and that the truth of this assumption is generally known, so that secondary and tertiary salience go together, and so on. He presents an analysis of game-theoretic reasoning which leads players to coordinate on the primarily-salient equilibrium. He then links this theoretical result with Schelling's evidence by proposing, as a separate empirical hypothesis, that conspicuous uniqueness and primary salience go together: as a last resort, people tend to pick the conspicuously unique.

Lewis's theory is very different from Bacharach's. In contrast, the next theory we consider, that of David Gauthier (1975), is a precursor of variable frame theory. Gauthier's central example is a coordination problem that is structurally similar to Heads and Tails. Two people are trying to meet. One is travelling to London by train from Leicester, the other is already in London. There are two places at which they could meet, Marylebone station and St. Pancras station (both of which have services to Leicester). Gauthier argues that rational players can solve their problem by redescribing it as a Hi-Lo game in which each player chooses between the options *seek salience* and *ignore salience*. These options are analogous with *choose the prominent* and *pick a thing* in the variable frame analysis of Heads and Tails in section 4 (see figure I.10). In Gauthier's story, St. Pancras is salient by virtue of having a more frequent train service to Leicester. (Gauthier is not assuming that, if the travelling player picks among trains from Leicester at random, he is more likely to end up at St. Pancras. Rather, Gauthier is interpreting the frequency of trains as a clue in Schelling's sense—a clue that is likely to come to mind for both players in the context of a problem that involves a train journey.)

Here is Gauthier's explanation of how rational players arrive at this particular redescription of the game:

> Our problem [as players] is that we have two equally good meeting places. What we need is a way to restructure our conception of the situation so that we are left with but one. We must restrict the possible actions which we consider, in such a way that we convert our representation of the situation [into a Hi-Lo game]. . . . We effect the necessary restriction by singling out some characteristic of some one equilibrium outcome. This characteristic then determines a new conception of the situation in which each person has but two courses of action—to seek an outcome with that characteristic, or to ignore that characteristic in what he does. (pp. 210–11)

The players need to coordinate on the same characteristic, but cannot communicate with each other. Having defined 'the salient' as 'that which is apprehended as standing out from the others', Gauthier concludes: 'Necessarily, therefore, the restricting characteristic is salience' (p. 211).

Notice that Gauthier's analysis differs from Bacharach's by treating the players as *choosing* between alternative descriptions of a game: each chooses the description which, if used by both, would lead to the best results for both. In contrast, the 'framed game' that Bacharach analyses just is the game as the players describe it to themselves. Thus, Gauthier eliminates the act-descriptions *choose the Marylebone station* and *choose the St. Pancras station* on the grounds that both players benefit by this elimination. In Bacharach's analysis of this game, *Marylebone station* and *St. Pancras station* are predicates which belong to the same family, and so the corresponding act-descriptions, though accessible to the players, are debarred as rational choices by the principle of insufficient reason. Bacharach thought that Gauthier's approach blurred the distinction between internal rationality (that is, acting on valid practical reasoning) and external rationality (that is, acting in a way that turns out to be to one's benefit). In his preliminary notes for chapter III, Bacharach writes: '[It] seems to me that one can't just go round changing one's own description for convenience; this is like changing beliefs; surely you must describe the world as you find it!'

Sugden (1991, 1993, 1995) developed a theory of focal points in parallel with Bacharach's development of variable frame theory. The two theories are not entirely independent: while developing them, Bacharach and Sugden exchanged ideas. Sugden's theory can be thought of as an attempt to formalise the intuition expressed in

Schelling's discussion of the game of 'Name a positive number'. Recall Schelling's remark that, if one asks what rule of selection would lead to unambiguous results, one may be struck with the fact that there is a first or smallest number, namely 1. Sugden proposes that players construe their problem as choosing among *rules of selection* rather than among objective strategies. Roughly, a rule of selection is a general principle, applicable to a broad class of games, which, when applied by a specific player to a specific game from this class, picks out a unique objective strategy. In the context of 'Name a——' games, rules of selection might include *name the first——that come to mind, name the most famous——, name your favourite——*, and *name the——which strikes you as most clearly unique*. In the context of a specific game, a rational player chooses to follow the *best rule*—the rule which, if followed by both players in that game, would yield the best results for both. In the case of 'Name a positive number', the rule with the most credible claim to be best is *name the number which strikes you as most clearly unique*. This is for the reason given by Schelling, namely that there happens to be one number which most players will perceive as the most clearly unique. In a different 'Name a——'game, a different rule might be best, but the principle of choice remains the same.

There are significant similarities between Sugden's theory and Bacharach's. Both work by redescribing the objective game in a way that transforms it into a Hi-Lo game. In Sugden's theory, the analogue of an act-description is a rule of selection; whichever rule is best is the analogue of *high*. Sugden's theory, however, focuses on a different kind of uncertainty than Bacharach's. While Bacharach takes account of the possibility that some predicates may not come to mind for a player, Sugden assumes that the same set of rules of selection comes to mind for everyone. But while Bacharach assumes that the extension of each predicate is known to everyone for whom that predicate comes to mind, Sugden takes account of imprecision in the application of a rule of selection. In Bacharach's theory, two players who opt for the same *choose the*——act-description will end up selecting the same objective strategy. In Sugden's theory, two players who use the same rule of selection may select different objective strategies. For example, if two players of the game 'Name a composer' use the *name the most famous——* rule, one might select Mozart and the other Beethoven.[20] In Bacharach's theory, if the predicate *most famous* comes to mind for both players, then it will identify the same composer for each of them.

Finally, we mention the theories proposed by Maarten Janssen (2001b) and André Casajus (2001).[21] These theories build on variable frame theory by offering more general analyses of the concept of isomorphism which motivates Bacharach's principle of insufficient reason. Casajus's theory is particularly general. It distinguishes between (what we have called) the objective game and the *labels* by which the players describe that game to themselves. Descriptions of strategies are modelled in terms of a set of *attributes*. Each attribute has a distinct label—for example, *colour* or *shape*. For each attribute, there is a set of labels of a different kind, which refer to alternative forms that the attribute can take—for example, the set of labels for *colour* might be {*red, yellow, green, blue*}. Each strategy for each player has an *n*-dimensional array of labels, where *n* is the number of attributes in the model. For example, if the attributes are *colour* and *shape*, a strategy might have the array of labels (*red, cube*). Casajus defines two games to be isomorphic if one can be transformed into the other by a one-to-one 'translation' of labels. He then requires that solution concepts should be invariant with respect to such translations. Casajus's attributes are similar to Bacharach's families of predicates. Casajus's invariance principle, however, is more powerful than Bacharach's principle of insufficient reason.

As an example, consider a blockmarking game in which there are three objects: a red cube, a blue cube and a blue pyramid. In variable frame theory, there are five act-descriptions: *pick a thing, pick a blue, pick a cube, choose the red* and *choose the pyramid*. None of these is debarred by the principle of insufficient reason, and so the framed game is not a Hi-Lo game. (There are two act-descriptions, namely *choose the red* and *choose the pyramid*, for each of which it is true that, if both players opt for it, coordination is guaranteed.) In Casajus's theory, each player chooses from the set {(*red, cube*), (*blue, cube*) and (*blue, pyramid*)}. Since *red* is symmetrical with *blue*, *cube* with *pyramid*, and *colour* with *shape*,[22] choosing (*red, cube*) from this set is isomorphic with choosing (*blue, pyramid*), while choosing (*blue, cube*) is isomorphic with neither of the other options. Thus we end up with a framed game with a Hi-Lo structure, in which (*blue, cube*) corresponds to *high*.

6. Envoi

We have now taken the reader to the point at which Bacharach's chapter IV would have begun. We have presented the main features of variable frame theory, and we have shown how this theory can be used to transform pure coordination games into Hi-Lo games. However, the 'first puzzle of game theory' has not yet been properly resolved, because we have not shown that it is rational to choose *high* in Hi-Lo. We now hand over to Bacharach.

Notes

1. If each player chooses *heads* with probability h and *tails* with probability $1 - h$, and if the probabilities for the two players are independent, the probability of their coordinating on the same label is $h^2 + (1 - h)^2$. If $h = 78/90$, as in Mehta, Starmer and Sugden's (1994) data, the probability of coordination is 0.77.

2. Sally (1995) reports a meta-analysis of 130 Prisoner's Dilemma experiments carried out between 1958 and 1992. Summing across the whole sample, the proportion of subjects choosing *cooperate* is 47.4 per cent (p. 62).

3. In order to conclude that (*cooperate, cooperate*) is the best pair of strategies for them, the players have to judge the payoff combinations (3,0) and (0,3) to be worse 'for them' than (2,2).

4. As we explained in the preface, Bacharach rejects this presumption.

5. In chapter 4 of this book, Bacharach presents evidence from an experiment he conducted with Gerardo Guerra. Further evidence comes from an as yet unpublished experiment, in which Nicholas Bardsley presented fifty-six Dutch students with two Hi-Lo games. In one game, the ratio of the money payoffs to *high* and *low* was 10:1; in the other it was 10:9. In each case, fifty-four subjects (96 per cent) chose *high*. Notice the qualification that the strategies have neutral labels. In section 7 of chapter 4, Bacharach reports an experimental investigation of Hi-Lo games with four labels, one of which was particularly salient, while not being associated with the highest payoff pair. This label was chosen by a substantial minority of participants. Notice also we are referring specifically to Hi-Lo games, and not to the more general set of games in which there are two or more Nash equilibria, one of which Pareto-dominates the others. There are games of this general type in which players tend to coordinate on Pareto-dominated equilibria; intuitively, the explanation is that the Pareto-dominant equilibrium is 'more risky' than the dominated one (Van Huyck, Battalio and Beil 1990).

6. For example, suppose that each player believes that his opponent is just as likely to choose one strategy as the other. Then both will choose *high*. Or suppose that each player believes that his opponent believes that he is just as likely to choose one strategy as the other. Then each player will expect his opponent to choose *high*, and so choose *high* as a best reply.

7. Bacharach's 'Scientific Synopsis' is a substantial document which he wrote in 2000 to describe his proposed book to potential publishers. It contains detailed synopses of thirteen planned chapters. Later he made considerable changes to the original plan, but the 'Scientific Synopsis' is one of the main sources of information about the intended content of the chapters that remained unwritten when he died. Unattributed quotations in section 1 of this introduction are from the synopsis of chapter I.

8. Within Bacharach's theoretical models, 'payoff' (or 'utility', which he treats as a synonym) is a primitive concept. He generally avoids making explicit statements about the interpretation of this concept. However, the ways in which he uses it seem to depend on the interpretation we have offered. This interpretation is implicit in the definition of 'validity' in the passage quoted above. The fact that he treats payoffs as utility indices in the sense of expected

utility theory, and his apparent endorsement of that theory as a component of practical reasoning (see the passage quoted at the beginning of this section), suggest that he sees his interpretation of 'payoff' as similar to that used in classical game theory. Since his analysis is directed at *finding out* what it is rational for an agent to choose, given her (and others') payoffs, he cannot interpret payoffs in the revealed preference sense, as descriptions of an agent's choices. To interpret payoffs as quantities of some specific 'material' good (such as money) seems inconsistent with Bacharach's aspirations to generality. In an 'offcut' from his notes for chapter VII, he refers to an argument which involves 'reading the payoffs [of a particular game], illegitimately, as amounts of money': the implication is that payoffs are *not* to be interpreted materially.

9. For a brief review of the relevant evidence, see Camerer (1995, section II).

10. This section is based on Bacharach (2003), the published notes of a 2001 lecture, from which the unattributed quotations are taken.

11. Here, for ease of exposition, we use a preliminary definition of 'frame'. Later we will define frames in a way that groups predicates into 'families'.

12. For example, see Bacharach (1998). See also Broome (1991, pp. 90–120), who argues that the theory of rational choice needs 'rational requirements of indifference'—that is, substantive principles which limit the fineness with which the world is cut up (or, in Broome's terminology, the fineness with which options are individuated).

13. Section 4 draws on Bacharach (1991, 1993, 2003) and Bacharach and Bernasconi (1997). As editors we face the problem that Bacharach's work contains no canonical statement of variable frame theory. The most formal presentation is in the working paper of 1991. The published paper of 1993 presents an abridged and nontechnical version of the theory, referring the reader to the working paper for more details. The 1997 paper is an experimental test of the theory, which summarises the theory only insofar as is necessary to explain the tests. The 2003 paper presents an analysis of frames, but not of decision-making in games. There are significant changes in assumptions and terminology among these four papers. A further problem is that Bacharach proposed variable frame theory before developing the theory of team reasoning. In variable frame theory, players reason as individuals. The choice of the profile (*high, high*) in Hi-Lo is treated in terms of equilibrium selection rather than of team reasoning. It seems clear, however, that Bacharach came to see the theory of team reasoning as providing the better explanation of this choice. See also note 14.

14. Variable frame theory does not allow players to choose arbitrary probability-mixes of act-descriptions. It is fundamental to the theory that acts can be chosen only 'under descriptions'. Thus a player cannot choose to randomise between two options unless that randomisation is itself an act description. One consequence of this feature of variable frame theory is that there can be games which have no variable frame equilibria.

15. The question of which of these two analyses is more appropriate has a close analogue in the theory of 'circumspect' team reasoning, when the 'we' frame does not have availability 1. Then the problem arises of how a player for whom the 'we' frame has come to mind should deal with the possibility that

his opponent has only the 'I' frame. Bacharach discusses this problem in section 3.3 of chapter 4 of this book, and argues that a player should keep to a team plan rather than revising that plan in the light of his private knowledge that he has the 'we' frame. This position is analogous with the analysis of Large and Small Cubes which implies the critical value $p = 1/2$. This 'no revisions' analysis of team reasoning first appears in Bacharach (1999, p. 126). In that paper, he thanks Sugden for pointing out a 'serious flaw' in an earlier version. That flaw was the analysis which allowed plans to be revised. One possible interpretation is that Bacharach changed his mind about how games such as Large and Small Cubes should be analysed in variable frame theory. An alternative interpretation (hinted at in Bacharach and Bernasconi 1997, note 7) is that the 'no revisions' analysis is proposed only in relation to the theory of team reasoning, and that Bacharach continued to endorse the earlier analysis in relation to variable frame theory, understood as a theory of *individual* reasoning about framed games.

16. Bacharach's analysis is in his 1991 paper. He considers a pure coordination game invented by Gauthier (1975), which we describe in section 5. We prefer to use Heads and Tails as an example because the tendency for people to coordinate on *heads* in this game is an established fact.

17. Sugden thinks he recollects Bacharach using this argument in support of his analysis of Schelling games in which each player has just two objective strategies.

18. In a recent conversation with Sugden, Schelling expressed some regret about bringing classical game theory into *The Strategy of Conflict*. In retrospect, he thought it might have been a better strategy to have presented his analysis of coordination as a contribution to interdisciplinary, empirically-based social science, orthogonal to the a priori, mathematical analysis of game theory.

19. The concepts of primary, secondary and higher-order salience are discussed by Mehta, Starmer and Sugden (1994). These authors contrast these concepts with what they call *Schelling salience*, which is based on the idea of 'finding the best rule', as earlier proposed by Sugden (1993) as an interpretation of Schelling (and explained later in this section).

20. The polar case of this form of imprecision is the case of *scrambled labels*, in which each player recognises the same predicates as the other but, for each predicate, its extension for one player is uncorrelated with its extension for the other. For example, consider a blockmarking game in which each player sees three apparently identical blocks arranged in a row—one on the *left*, one in the *centre*, one on the *right*—but in which the arrangement of the blocks is randomised separately for each player; they are rewarded for marking the same 'objective' block. Crawford and Haller (1990) analyse this case in relation to repeated coordination games.

21. Janssen was one of the first game theorists to develop Bacharach's ideas on framing. This 2001 paper appeared as a discussion paper in 1996.

22. Strictly, this is true in Casajus's theory only if the two sets of attributes have the same number of elements (Casajus 2001, pp. 42–43).

Chapter 1

The Hi-Lo Paradox

1. The Game of Hi-Lo

You and another person have to choose whether to click on A or B. If you both click on A you will both receive £100, if you both click on B you will both receive £1, and if you click on different letters you will receive nothing. What should you do?

It is obvious that the only rational choice is to click on A. Yet oddly, game theory has no explanation of *what makes* A-choices rational.

This is an example of the game of Hi-Lo. In the general case of Hi-Lo, each of n players chooses one item from the same finite set of alternatives without consultation. With each alternative goes a prize, and one alternative's prize is greater than all the others. If all choose the same alternative all get the prize that goes with it, and if not everyone chooses the same alternative none gets anything. Figure 1.1 shows the payoff matrix of a Hi-Lo in which there are two players, 1 and 2, and two alternatives, A and B, with associated prizes of 5 and 1.

You are to play Hi-Lo, and it is common knowledge that you and your coplayer are intelligent people. It seems quite obvious that you should choose A. However, the question why it seems obvious, and the related question of why people almost always do choose A, have turned out to be anything but easy to answer. This chapter explores these questions. The answer I shall offer has far-reaching implications for game theory and, I shall argue, for our conception of ourselves as social beings.

		Player 2	
		A	*B*
Player 1	*A*	5, 5	0, 0
	B	0, 0	1, 1

Figure 1.1. Hi-Lo

It may be thought that such situations are artificial and rare, that 'we should be so lucky!' But Hi-Los are very prevalent in our lives. It is just that the Hi-Lo structure of the payoffs is often not transparent. I first give examples of Hi-Los, of varying transparency, in varied human activities. Next I describe what is known about the choices people make. As we might expect, the evidence is that in cases in which the structure is transparent, people are overwhelmingly successful in coordinating on (A,A), and that we have great facility in Hi-Lo. It appears, too, that there is a strong intuition that choosing A is the only rational thing to do.

Example 1. Running a single. In the game of cricket two batsmen—the striker and nonstriker—stand at two 'wickets', one at each end of a twenty-two-yard strip. The striker tries to hit a ball projected towards his end of the pitch far enough that each batsman can get to the wicket at the other end before the other side retrieves the ball and strikes one of the wickets with it, in which case one of the batsmen is said to be 'run out', a serious setback for his side. If the batsmen succeed their side's score goes up by one 'run'. If they don't run, both sides' scores are unaltered. If one runs and the other doesn't, it's likely that the runner will find himself run out. If the striker hits the ball only a short distance, then even if both run, one of them is likely to be run out, but if he makes a good hit and they both run, they are likely to add one run to their side's score. Suppose the hit is good enough. Each batsman may either run or stay; so the payoff matrix has the form of figure 1.2.

Example 2. Focal coordination. As we have seen, Schelling games in option form are typically Hi-Lo. [MB is referring to material that would have been in the unwritten chapter III. The material is discussed by the editors in section 4 of the introduction.]

Example 3. Who fetches which? Lizzie and I fail to talk in the morning about who is going to fetch Julian from his school and who will fetch Emily from hers; and it's now too late to get in touch. It is common knowledge between us that I have a meeting near Emily's school, that she will be near Julian's, and that we share the objectives of fetching

		Nonstriker	
		run	*stay*
Striker	*run*	1, 1	–20, –20
	stay	–20, –20	0, 0

Figure 1.2. Running a single

		Me: *fetch Emily*	Me: *fetch Julian*
Lizzie	*fetch Julian*	2, 2	−2, −2
	fetch Emily	−2, −2	1, 1

Figure 1.3. Who fetches which?

		Player 2: *wide*	Player 2: *stay*
Player 1	*pass*	5, 5	0, 0
	shoot	0, 0	1, 1

Figure 1.4. Vision

both children on time and minimizing total time spent fetching. So we face the problem in figure 1.3.

Example 4. Vision. At a certain moment in a football match, Player 1 has the ball, and the obvious move is to shoot, but defenders are blocking his path. The chances of finding the net are therefore only slight, but if Player 2 stays where he is he may well keep possession if the ball rebounds. There is also another option. If Player 2 runs wide, a pass from Player 1 to Player 2 would allow Player 1 to find space in front of goal, Player 2 to make a return pass and Player 1 to have a good sight of goal. Player 1 and Player 2 face the problem of figure 1.4.

In Example 1 the options are obvious—run or stay. But here they are not. It takes what footballers call 'vision' for Player 1 to be aware of the pass-wide option for his side, and either 'vision' or 'telepathy' for Player 2 to see his part in the move, if there is no chance for Player 1 to attract his attention. Decision problems are often, as here, not given exogenously in their entirety but are in part made—by the creative perception of possible options by the decision-maker.

Example 5. Telling the truth. An act utilitarian is someone who in each choice situation in which she finds herself chooses with the aim of maximizing U, the sum of the utilities of all individuals in some reference group. If the outcome of her choice depends on the choices of certain others, the decision situation is a game, and if she and these others are all act utilitarians, and there is complete information, it is a pure

		Player 2	
		B	*D*
Player 1	*T*	2, 2	−2, −2
	F	−2, −2	2, 2

Figure 1.5a. Truth telling: With strong disbelief

		Player 2	
		B	*D*
Player 1	*T*	2, 2	0, 0
	F	−2, −2	0, 0

Figure 1.5b. Truth telling: With weak disbelief

coordination game. Hodgson (1967) argues that in deciding whether to tell the truth, act utilitarians confront a coordination game of the form of a simple Schelling game. The speaker Player 1 can tell the truth (T) or a falsehood (F); the hearer Player 2 can believe (B) or disbelieve (D). It is assumed that U is high (2, say) if the truth is believed and low (-2, say) if a falsehood is believed. In some cases believing the truth and disbelieving the falsehood have exactly the same effects (they do if the truth is a proposition P, the falsehood is $\neg P$, and disbelieving means believing the negation), and so do believing the falsehood and disbelieving the truth, so the payoffs are those of figure 1.5a. Hodgson argues that act utilitarians are left not knowing whether or not to tell the truth, and rejects this form of utilitarianism because of this indeterminacy. Gauthier (1975) notes, however, that if a disbeliever merely fails to form a belief at all, rather than believing the opposite of what is asserted, then the outcome of (F,D) is worse than that of (T,B). In this case the matrix is that of figure 1.5b, a (weak) Hi-Lo. On Gauthier's interpretation of disbelief, if A is the unique rational solution of Hi-Lo then act utilitarianism gives clear advice about truth telling. As to whether A is the unique rational solution, that is the topic of this chapter.

Example 6. Calling a catch. A second Hi-Lo drawn from cricket illustrates a fundamental form of organization. One basic function of organization is to amplify the information on which individual actions are based. This allows a profile of actions that is better geared to the state of the world than any of those attainable by the players without organization. If the procedural cost of the organization is not too great, the enlarged game in which players have the option of conforming to this profile is a Hi-Lo in which conforming to it is A.

When a batsman hits the ball in the air, if a fielder catches it before it touches the ground the batsman is 'caught out', which gives a significant boost to the fielding side (a payoff of 20). If the ball is 'skied', catching it means tracking its flight while running towards it. This is a difficult task which absorbs all the fielder's attention. If two fielders (F1 and F2) are near and both go for the catch there may be a collision, and the likely upshot of this is that the batsmen score two runs or so, and a fielder may be injured (a payoff of −5). If neither goes for the ball the likely upshot is two runs or so (a payoff of −2).

Typically each of the fielders will have some idea of whether he or the other is better placed to make the catch, but can be wrong. We model this by supposing that there are two equiprobable states of the world, ω_1 and ω_2. In ω_1, F1 is better placed and will make the catch with high probability (1.0) if he is unhindered, and F2 would make it with a lower probability (0.75). In ω_2 the players' positions are reversed. Whatever the state, each player gets right which state it is with probability 2/3 and gets it wrong with probability 1/3.

To begin with, I suppose that each fielder has two options: catch if you think yourself better-placed (B), and catch if you think yourself worse-placed (W). This is a simple Hi-Lo, in which each of these is in Nash equilibrium with itself (figure 1.6a). (The mathematical derivation of the payoffs is given in the appendix to this chapter.)

Some cricketing teams use the following calling routine. The captain calls the name of the fielder he thinks has the better chance, and the captain is always right about this. A fielder therefore has a third option: obey the call (C). But C carries a procedural cost: attending to the call takes precious seconds and reduces the chance of success if you do end up going for the catch—if unhindered the better-placed has probability of 0.5 and the worse-placed 0.375. The upshot is

		F2	
		B	*W*
F1	*B*	9, 9	5.8, 5.8
	W	5.8, 5.8	7.3, 7.3

Figure 1.6a. Calling a catch: Without organization

		Player 2		
		C	*B*	*W*
	C	10, 10	8.8, 8.8	2.7, 2.7
Player 1	*B*	8.8, 8.8	9, 9	5.8, 5.8
	W	2.7, 2.7	5.8, 5.8	7.3, 7.3

Figure 1.6b. Calling a catch: With organization

shown in figure 1.6b. (Again, the mathematical derivation of the payoffs is given in the appendix to this chapter.)

The situation is a Hi-Lo in which (B,B) and (W,W) are still equilibria, but better even than (B,B) is (C,C): the extra information more than compensates for the procedural cost of (C,C).

Example 7. Deciding together. Suppose that two people have exactly the same aims, and have the opportunity to decide together what to do. Suppose these common aims are all the aims either has that could be affected by their decision. Suppose the problem they jointly face is a matching game with options A, B and C. For example, a colleague and I might have the opportunity to decide together whether to take a visiting speaker to lunch at restaurant A, B or C, and are concerned only to eat well, within budget and in time for the seminar. Deciding together is exchanging information, weighing considerations, agreeing on the judgement that a certain alternative is best and deciding to adopt that alternative. If the two people decide together, then at the stage of this process at which they reach agreement on a judgement they face a Hi-Lo.

Example 8. Calling a run. Example 1 was about the situation in cricket in which it is clear to both batsmen that a single run is 'on'. But in other cases this may require judgement, and the two batsmen may not be equally well placed to make it. Some cricketers use the following practice for deciding whether to run. If the striker hits the ball 'forward of square', so that it subtends an angle less than 90 degrees with the line of the pitch, he calls Yes or No; if he hits the ball backward of square, the nonstriker calls. Both run if and only if the call is Yes. Call this the Variable Caller practice. It puts the decision in the hands of the batsman better placed to make a good judgement quickly. To conform with the Variable Caller practice, a batsman implements the following strategy (V): if you hit the ball forward of square, or your partner hits it backward of square, call Yes if you think a run is on, otherwise No; if you hit it backward, or your partner hits it forward, do not call; obey your partner's call; if he fails to call, stay. Suppose, simplifying, that to their common knowledge the two options each batsman considers are Run if you think there's time, and Variable Caller (V). Then the situation is a Hi-Lo in which (V,V) is the efficient pair. This example resembles Example 6 except that the judgement controlling the basic action each player should take—taking a catch or staying in Example 6, here running or staying—was there made by a third party and here made by one of the players themselves. It is an example of a self-organizing team, and an archetype of a very large class of organizations.

Example 9. Unlocking. Figure 1.7 shows a situation of a type that was common in the town in which I live during the 'manhole years' that followed the privatization of public utilities and the laying of cable. I myself was in car 3. In the figure the diagonal arrows represent the direction signals cars 1, 2 and 4 were making; the horizontal arrows mean that cars 3 and 5 were making no direction signals. The five cars all came to a halt, everyone in view of everyone. A few seconds later we started to move: first, car 2 turned right into Abingdon Road, car 1 remaining stationary; next, car 4 turned right into Whitehouse Road

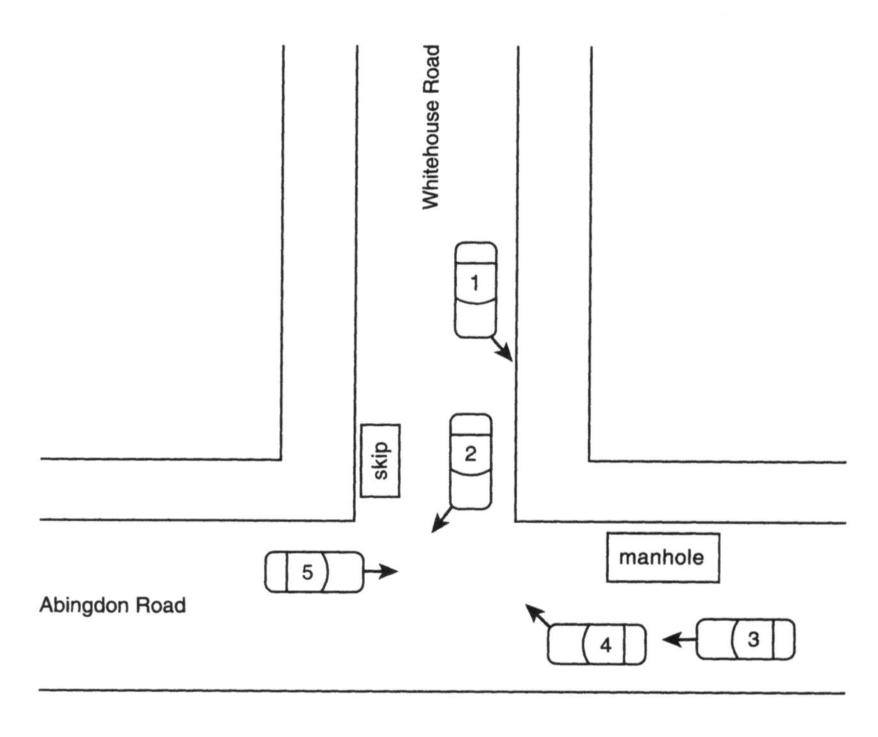

Figure 1.7. Unlocking

and threaded between the skip (dumpster) and car 1; next, car 3, my own, went straight on down Abingdon Road; next, car 5 went straight on up it; finally, car 1 passed the skip and turned left. Assuming that everyone wanted to be on his or her way in as little time as possible, this traffic unlocking problem is a Hi-Lo. The unlocking process that took place took perhaps 45 seconds. There were other possible combinations of movements, one involving car 4 forcing cars 1 and 2 to reverse. All were likely to take longer for each of the drivers. The five of us solved the Hi-Lo in a matter of seconds.

These examples begin to show the importance of situations of Hi-Lo form in our lives. They arise whenever there is a common purpose and a best method of furthering it, and mixing methods is worst of all. Game theory has made us subtly sensitive to conflicts of private motive, not least when people subscribe publicly to a common aim; but this should not blind us to the huge domain of decisions we make in which aims are exactly or approximately the same. This does not require that they be stably so, only that there are occurrent frames in which they are. And this is normally the case when we carry out the normal activities of members of teams and other organizations to which we belong. Within the big game in which we may contemplate our department or firm or family as an entity outside ourselves, with whose goals our own purposes may conflict, we daily play subgames in which we frame ourselves as functioning members of some such system. When we do we are usually playing Hi-Lo.

2. The Data

There are these two broad empirical facts about Hi-Lo games: people almost always choose A, and people with common knowledge of each other's rationality think it obviously rational to play A. Call these the *behavioural fact* and the *judgemental fact* about Hi-Lo.

In cricket, when it is clear that there is a safe run to be taken, and the payoffs are therefore as in figure 1.2,[1] it is unheard of not to take the run. In Schelling games, the only formal explanation we have of the coordination success we observe—variable frame theory—depends on the assumption that people play A in the induced Hi-Lo. In these examples, people appear to be disposed to make A-choices even though the Hi-Lo element in their problem may not be transparent. Their problem is not displayed to them as a Hi-Lo, but we can still explain their behaviour as resulting from A-choices in an implicit Hi-Lo; moreover, if they do implicitly confront Hi-Los, their behaviour entails that they make A-choices in them. In the Hi-Los which emerge when we decide together which alternative is best, as in Example 7, to agree, genuinely, that restaurant A is best and yet not to decide to go there seems cockeyed. It seems as absurd as concluding by yourself that a certain alternative would be best for all the aims that weigh with you, and then not choosing that alternative.

When it comes to laboratory tasks or text-book examples in which the form is transparent, the evidence that people are inclined to choose A is extremely commanding. Consider a Hi-Lo game 1, presented to two players for money prizes of £10 and £1. It seems so obvious that everyone will choose A that there were apparently no experimental tests before 2001 of the hypothesis that people choose A in overt Hi-Lo games. Then, in the course of an experimental programme to which I shall return, Gerardo Guerra and I asked each subject in a pair to choose one of the cards in a display (figure 1.8), in which the prize for matching on each of the items is shown below it. Although the differences among prizes were here quite small, fifty-eight subjects out of sixty-four, or 91 per cent, chose the £6 card.[2]

Most of us appear to have an overwhelming intuition that it is rational to play A, and moreover that this is obvious. The conviction seems to be independent of how rational we think our coplayer: in particular, it

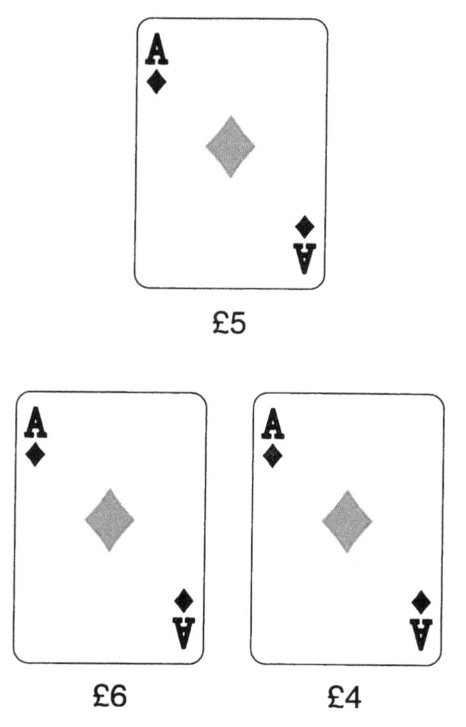

Figure 1.8. Choose a card

applies just as much when we think our coplayer highly rational as when we think he is simple-minded. Moreover, it applies just as much when there is common knowledge of the rationality of us both. This is what I have dubbed the judgemental fact about Hi-Lo. The judgemental fact is illustrated by the widespread endorsement of the intuition by theorists considering play between symmetrically rational players. Several assert or imply that A is obviously rational (Lewis 1969; Gauthier 1975; Farrell 1988; Harsanyi and Selten 1988; Anderlini 1999; Crawford and Haller 1990; Bacharach 1991; Sugden 1995; Janssen 2001a).[3]

One might expect the explanation of these two facts about Hi-Lo, the behavioural and the judgemental fact, to be straightforward. But these facts prove to be mysterious, and their explanation anything but that.

3. The Hi-Lo Paradox

3.1 What Game Theory Predicts

The basic solution concept of standard game theory, Nash equilibrium, delivers two solutions to Hi-Lo: (A,A) and (B,B). (B,B) is just as good a solution as (A,A), for it satisfies just as well the logic of Nash equilibrium, namely that beliefs plus instrumental rationality yield facts which match the beliefs. *If* I think you will choose B then the act it's best for me to choose is B, and *if* you think I will choose B then it's best for you to choose B, so both our hypothesized expectations are correct. (B,B) expectations get confirmed by best-replying agents.

Because (A,A) and (B,B) are both Nash equilibria, the Nash equilibrium solution concept predicts little or nothing. Whether little or nothing depends on how we interpret the claim that these equilibria are solutions. The strong form of the claim is that there is some equilibrium that all rational players (who are commonly known to be rational) will play. The weak form is that every such player will play her component in some equilibrium. The strong form predicts little, and the weak form nothing. In either form, game theory fails to explain why everyone chooses A.

3.2 The Paradox

The predictive failure is bad news for game theory in two different ways. First, because game theory aspires to explain actual behaviour, and here fails to explain a particularly strong regularity of behaviour. But second, because game theory is a theory of rational choice and so should deliver solutions which accord with, not actual behaviour, but another sort of empirical data, namely our intuitions about what is rational. To be sure, our intuitions are both partial and unreliable evidence about the rationality of an action. We should not expect people to be able to identify the reasoning principles that govern their conclusions, even when these principles are sound: for example, most people easily and reliably reason in accordance with *modus ponens*, but almost no-one can tell you that it is *modus ponens* that sanctions their conclusions. We know something makes sense (playing A, par excellence); but we do not know what sense it makes. And our intuitions may not be reliable: we may be deceived in finding some theoretical or practical proposition persuasive; one sort of evidence for this is that on reflection it loses its initial aura of obvious rightness. But the more stable under reflection our intuitions are, the stronger they are, and the more they are shared by serious thinkers about the subject matter, the greater is the challenge to theory to encompass them.

A paradox is a mismatch between such high-quality intuitions and the deliverances of an accepted theory. The clash between the obvious rationality of the choice A and the inability of game theory to single out A is a paradox. It is a weak paradox, because game theory does not predict that A will not be chosen. It is a paradox nonetheless, a failure of theory to agree with intuition.

3.3 *Refining Equilibrium*

Such paradoxes are not uncommon in game theory. Some have been dealt with in the past by 'refinement' of the Nash equilibrium solution concept. The broad object of the equilibrium refinement programme of the 1970s and 1980s was to eliminate equilibria which were intuitively suspect, by adding to the Nash solution principle further rationality conditions on players' choices of a decision-theoretic character. To eliminate in this way a suspect Nash equilibrium (A,X), we seek a decision-theoretic argument that A is not, on closer examination, or on a more careful specification of the game, a rational choice against the choice X.

A simple example is the elimination of Nash equilibria in which one act is weakly dominated. Suppose the payoffs are as in figure 1.9.

		Player 2	
		X	*Y*
Player 1	*A*	1, 6	1, 3
	B	1, 4	2, 5

Figure 1.9. A game with a weakly dominated Nash equilibrium

(A,X) is a Nash equilibrium (A is *a* best reply to X), but A is weakly dominated by B since B is at least as good as A against all feasible acts of Player 2, and better against one, Y. Now it is argued (Selten 1975) that even rational players, having made a choice, may with nonzero probability do something else by accident. In the present case, consider a state in which Player 1 is certain that Player 2 will choose X and Player 2 is certain that Player 1 will choose A. Then Player 2 will indeed choose X, but there is some probability ε that her 'hand will tremble' and she will *play* Y. This means that the expected utility of A for Player 1 is lower than that of B, so it is not rational for Player 1 to choose A. The weak-dominance pattern in A's payoffs means that this is true even for infinitesimal ε. Selten concludes that to be a rational solution, a Nash equilibrium must not include a weakly-dominated act.

Neither this requirement nor any other refinement which has commanded general acceptance has the effect of eliminating (B,B) in Hi-Lo. Two refinements proposed by Harsanyi and Selten (1988), called *risk dominance* and *payoff dominance,* do eliminate (B,B). As we shall see in a minute, however, risk dominance is suited to describe the choices of players whose rationality is bounded in a certain way, but not to players with common knowledge of full rationality. Payoff dominance is another matter. Its logic is closely related to the main thesis of this book. I shall argue that in some circumstances the outcomes it selects will tend to be generated by the deliberations of fully rational agents. But the rational considerations that generate it are quite different from the decision-theoretic ones that the equilibrium refinement school was searching for. The conclusion: the Nash solution principle implemented using the most stringent version of standard rationality fails to deliver A. (B,B) is not only an equilibrium; it is a perfectly good equilibrium. But A is obviously the unique rational choice: this is the paradox.

It may be protested that there is no real paradox here because the game-theoretic argument for B is flawed. It first supposes that each player expects B, then proceeds to show that these expectations are part of a consistent constellation of beliefs and acts. But it does nothing to show where the expectation might come from in the first place. Indeed, such an expectation is intrinsically implausible, so it postulates the intrinsically implausible. How could anyone expect a rational player to play B?

Indeed it seems that no-one could. But the problem is to show why—why rationality excludes B.

It must be understood that game theory does not predict B; it merely fails to exclude B. There is nothing fallacious about that, only something incomplete. The programme game theory sets itself is to plot all outcomes *consistent with* all its postulates about the players' reasoning and knowledge. Ultimately, the reason why (B,B) is a solution is that it is consistent with all the facts about rationality that game theory can muster. (B,B) is a solution because game theory has mustered no fact about rationality that excludes B. Given this absence, the rigorous project of game theory obliges it to allow belief in B, and then B as a response to this belief. Given the absence, so far from criticizing game theory for allowing B and the equilibrium (B,B), we should praise it for its clear-headed acknowledgement of the consequences of its own limits. We should only regret that our current formal theory of interactive behaviour has a serious gap, no response to the question 'What's wrong with B?' or, equivalently, 'What's right with A?'

4. The Response

Since game theory fails to explain why people would choose A in Hi-Lo, any explanation of A-choices must say they are not the result of game-theoretic reasoning in Hi-Lo, or at least not of this alone. Game theorists, philosophers, psychologists and others have sought to explain A-choices. The theories advanced have fallen into the same two broad categories as theories that have tried to explain C-choices in the Prisoner's Dilemma: respecification theories and bounded rationality theories. [Throughout the text MB uses the standard notation of 'C' and 'D' for the strategies "Cooperate" and "Defect" in the Prisoner's Dilemma.] Respecification theories explain behaviour usually thought of as an A-choice in a Hi-Lo by saying that the chooser is not in fact playing Hi-Lo but some related game G in which game theory does predict A. Bounded rationality theories explain A-choices in terms of limits on or lapses in rationality. In the Prisoner's Dilemma literature we find respecification theories in which G is, for example, an indefinite repetition of the Prisoner's Dilemma, or a Prisoner's Dilemma with transformed payoffs, and bounded rationality theories in which players have limited depth or use magical reasoning.

There is a third sort of theory, which may be called revisionary; revisionary theories aspire to explain the target behaviour (here, A) as rational, but rewrite accepted doctrines of what is rational. Revisionary theories may add new principles of rational choice, like equilibrium selection theories, or extend the framework by adding new primitives, like variable frame theory, or challenge previously accepted canons, as some evidentialists deny the instrumentalist canons. The boundary between revisionary theories and bounded theories may be contentious even in the first two cases, for the extension championed by the revisionist may be seen by others as involving invalid reasoning. The theory advanced in this book is a revisionary theory which is not immune from the possibility of such a reaction.

5. Respecification Theories

Farrell (1988) and Anderlini (1999) remodel the game in which A is chosen as a game in which players can communicate with each other before choosing between A and B. Anderlini suggests that our intuition that it's rational to play A is due to an implicit assumption that players will manage to communicate their intentions. Farrell introduces a new rationality postulate (a refinement, called 'neologism-proofness') which applies in such games. The key idea is that some claims are 'self-signalling'—the speaker would like them to be believed if and only if they're true; Farrell postulates that it's rational to believe a self-signalling claim. Clearly both 'I'll play A' and 'I'll play B' are self-signalling. Intuitively, in Hi-Lo this makes it rational for the sender to send the message 'A'.

Aumann and Sorin's (1989) respecification theory is actually not for Hi-Lo but for a broader class of games, common-interest games, which includes Hi-Lo games as a subclass. A common-interest game is any in which there is a profile that is Pareto-superior to all others. A common-interest game need not be a coordination game. An example is Rousseau's Stag Hunt, shown in figure 1.10. [The story behind the game is that the two players are individuals living in a presocial 'state of nature'. Each chooses independently whether to hunt for rabbits (R) or deer (S, for 'stag'). Deer hunting requires concerted action by both individuals, while either can hunt rabbits on his own. Both individuals do better by hunting deer together than by hunting rabbits separately, but hunting deer alone is the least productive activity of all.]

		Player 2	
		S	*R*
Player 1	*S*	2, 2	–1, 1
	R	1, –1	1, 1

Figure 1.10. The Stag Hunt

If a common-interest game is a coordination game and symmetric, it is a Hi-Lo, and (A,A) is the unique Pareto-optimal profile of the definition. If it can be shown that in common-interest games in general the Pareto-optimal profile is played, it is shown a fortiori that (A,A) is played in Hi-Lo. Aumann and Sorin consider an arbitrary common-interest game and suppose that it is repeated indefinitely, and that with some chance each player has 'bounded recall'—she remembers past play only *n* rounds back. They show conditions under which, in this setting, the probability that a player plays A tends to 1 over time.

Both these models explain A-choices in a Hi-Lo game by interpreting this game as the kernel of a larger game. The phenomena they study are of intrinsic interest, but they only help explain the A-choices we are interested in (and the feelings people have about them) if the situations in which these A-choices are made really do feature communication or repetition. Looking again at the examples of section 1 it is clear that most of them do not actually have these features. In particular, one-shot laboratory experiments with subjects behind screens certainly do not. Is it not possible, though, that even if they do not, nonetheless the players treat them as if they did in the sense that they use heuristics appropriate to cases in which they do, perhaps because they are used to such cases? But this 'assimilation' hypothesis is of no help in the repetition case, because (A,A) is predicted only in the nonexistent long run; the model does not predict that the prospect of this long run encourages the use of A now.

In the communication case assimilation cannot be excluded out of hand. It is not inconceivable that, in deliberating in some Hi-Los, a player asks herself what message her coplayer would send if she were able, and best-replies to this counterfactual, imagined message. This hypothesis, however, is not advanced by Farrell and Anderlini themselves. It is also empirically implausible in some paradigm cases, and fails to satisfy the generality aim. In some Hi-Los, such as those that arise in fast-moving games, quick decisions are needed, and the roundaboutness of the reasoning on this imagined-message hypothesis makes it implausible. It doesn't ring true that a footballer wonders, even implicitly, what his teammate would signal in order to further his interest in winning the game if he had the chance to signal. And in cases in which one of the actions in the efficient profile is itself a message, as in Calling a Run, and in the vast array of organizational Hi-Los of which this is the archetype, the hypothesis means that players have thoughts about quite bizarre matters: the receiver deliberates that the sender would, if she had the chance, send a message saying that she was going to send a message, an amusing but improbable possibility.

6. Bounded Rationality Theories

In the literature that seeks to explain the behavioural fact—A-choices—we find two kinds of cognitive shortcomings: magical thinking and depth limits. I first describe what these are, then how they might explain A.

6.1 Magical Thinking

An agent's reasoning is called *magical* (Elster 1989, pp. 195–202) if it fallaciously attributes causal powers to her own decisions. Decision theory and common sense both hold that one sort of reason that can rationalize an action x is a *causal* reason, namely that doing x would *bring about* a better end-state than its alternatives would.[4] Someone who thinks magically claims to have reasons of this kind, and this much of her thinking is sound; her mistake is in her view of what causes what.

The specific mistake she makes in so-called evidential or diagnostic reasoning comes from the fact that *what she decides* is good evidence for something else. The decision-maker now slides from the idea that her choice is a good sign of the other's choice to the idea that it is a cause of it. Very often, perhaps typically, in our decision-making lives, what makes a choice x be evidence for an outcome o is precisely that x tends to cause o, and it seems likely that it is the normality of this case that lures us into the slide. But although it's normal it's far from guaranteed, so sliding is unsafe. Consider the case of smoking and cancer. Suppose you hear on the news that the conditional probability of contracting cancer is higher given that you smoke than it is given that you give up smoking: $P(\text{cancer} \mid \text{smoker}) > P(\text{cancer} \mid \neg\text{smoker})$. It is natural to conclude that giving up is a good idea, and what lies behind this is the presumption that the statistical relationship comes from a causal link from smoking to getting cancer. But this presumption might be false: it might be that the relationship comes from a common cause, for example, that the tendency to smoke and susceptibility to cancer are both expressions of a certain gene.

In the game case a player's choice of x can be evidence that another player will choose y. But games are by definition situations in which different players' acts are causally independent. This part of the definition of a game is due to the fact that game theorists want to model decisions made without physical communication, and assume that there is no such thing as telepathy. So if Player 1 believes, rightly or wrongly, that her choosing x is evidence that Player 2 will choose y, believing that this evidence is due to a causal link must be a mistake.

There is more than one scenario in which Player 1's choice can be evidence of Player 2's. The simplest is the case of similar players. If

Player 1 knows that she and Player 2 are 'like-reasoning', she can argue: 'Whatever choice x I make, Player 2 will make the same one; so I should choose x so that (x,x) is as good as possible.' This argument 'masks off' nonsymmetric pairs of choices and leaves Player 1 just scanning the main diagonal. In the Prisoner's Dilemma it leads a player to choose C, and in Hi-Lo to choose A. It is fallacious, because choosing C will make no difference to whether Player 2 chooses C or D (Lewis 1979). What is true is that if at the end of deliberation Player 1 has chosen C, it is probable that Player 2 will have too. But this fact may not be legitimately used by Player 1 during a deliberation not yet completed.

Bacharach and Colman (1997) propose a way in which Player 1 can see her choice as evidence of Player 2's in which Player 1's model of Player 2 is slightly richer, containing a best-reply step. Player 1 argues: 'Whatever I choose, Player 2 will "read my mind", and then do the best thing for himself given my choice.' This is a masking move which lets Player 1 confine her attention not to diagonal points (x,x) but to the points $(x, B(x))$, where $B(x)$ is Player 2's best reply to x, which is assumed unique. Colman and Bacharach suggest that a player goes on with the piece of magical thinking: 'So I should choose the x that makes the pair $(x, B(x))$ as good as possible'. The choice which it delivers is called the Stackelberg act. But this is not the end of the story. The Stackelberg act, say x^*, need not be a best reply against the expected act of the opponent $B(x^*)$—that is, $(x^*, B(x^*))$ need not be an equilibrium. It seems likely that in these cases the player may reject the Stackelberg act on reflection.[5] Colman and Bacharach conjecture that a Stackelberg act will tend to be chosen when but only when it has the equilibrium property. Colman and Stirk (1998) found evidence of this in a set of twelve 2×2 games, nine having and three not having the Stackelberg equilibrium property.

In Hi-Lo, A is the Stackelberg act and has the equilibrium property, so both these magical reasoning theories, the similar-player theory and the Stackelberg theory, explain the behavioural fact.[6] Both are bounded rationality theories, because magical reasoning involves a mistake.

Another explanation of A-choice, by Jacobsen (2001b), subtle and imaginative, claims to explain the choice as rational, but also seems to involve a hidden evidentialist element. In the Jacobsen theory each player selects a 'plan'. Player 1 selects the pair (x,y); x is her choice in her actual role (as Player 1), y is her choice for the act she would perform if she were to be Player 2. This seems to be another evidentialist theory. Think of me, choosing between plan (x,y) and (x',y'). According to Jacobsen, by the axiom Janssen (2001a) calls Internal Consistency, if I choose (x',y') rather than (x,y) this implies that my expectation of your act is y' rather than y. The reason given is that my expectation of what you will do depends on the corresponding part of my plan, that is, on what I decide I would do if in the other role. It is indeed reasonable for me to take this as evidence about what you will do. But since I cannot causally affect what you will do, I cannot use this epistemic consequence of my deciding on y' rather than y as a reason to change from x to x'. My changing from y to y', though evidence that you will do y' and evidence against the event that you will do y, does nothing to bring it about that you will do y' and not do y.

6.2 Depth Limits

An old idea for explaining A-choices is 'equiprobability'. The thought is that if Player 1 has no idea what Player 2 will do, then it's better for Player 1 to do A. For if she has no idea, then her personal probabilities for Player 2's doing A and B must be equal, and then A gives her a higher expected payoff (2.5 instead of 0.5 in the figure 1.1 example). This model is *asymmetric*: Player 1 treats Player 2 as unlike herself. For if Player 2 were like her in terms of his initial information and reasoning powers, he would presumably go through the same reasoning and would also choose A. So Player 1 cannot halfway rationally both think Player 2 is just like her and have no idea what Player 2 will do.

Harsanyi and Selten's principle of risk dominance turns out to boil down to the equiprobability principle—and so in my view to be inappropriate as an unbounded rationality solution concept. So I shall classify it as a bounded rationality explanation of A, even though that may not be how Harsanyi and Selten intended it. Harsanyi and Selten's general theory of rational solutions of games adopts the postulate that solutions must be Nash equilibria, then seeks 'refinements' of Nash equilibrium to deal with cases when there is more than one unrefined Nash equilibrium. Whereas the equiprobability argument is essentially decision-theoretic and selects an act directly, the risk-dominance theory first selects an equilibrium and then predicts that each player will play her part in it. For the class of symmetric 2×2 games, which includes Hi-Lo, Stag Hunt and the Prisoner's Dilemma, the risk-dominance principle is that an equilibrium is a solution only if it is not 'risk-dominated'; where, if E and E' are any two equilibria, E *risk-dominates* E' just if adherents of E do better than adherents of E' against coplayers equally likely to be adherents of either. If there are two equilibria with no acts in common, like (A,A) and (B,B) in Hi-Lo, or (S,S) and (R,R) in Stag Hunt, the risk-dominance principle gives the same act-choices as the equiprobability argument.[7] In Hi-Lo, risk dominance yields A.

Harsanyi and Selten themselves say little about why we should expect people to adhere to E only if doing so gives higher payoffs against an equal-probability mix of E and E' adherents. In evolutionary game theory, risk-dominance ideas arise naturally in studying the emergence of one equilibrium rather than another; the processes studied involve a hard-wired E population invaded by random waves of hard-wired E'-adherents, and what counts is the relative objective success of E-adherence and E'-adherence against the resulting probability mix of E and E'.

But here we are interested not in hard-wired but in deliberating agents. The probabilities in the mix are probabilities in the head of Player 1, specifically elements of Player 1's model of Player 2.

We can see better what is involved in the equiprobability theory by casting it in the framework of Stahl and Wilson's 'level *n*' theory of games (1995). This is a bounded rationality theory of how real players play games quite generally. In it, players reason strategically at different 'levels'. A level 0 player has no model of her coplayer; level 1 players believe they are playing level 0 players, and maximize expected payoff on this belief; level 2 players think a coplayer is level 0 or level 1 with probabilities adding to 1, and maximize expected payoff; and so on. Further, level 0 players are assumed to pick an option at random. Thus level 1 players of Hi-Lo are led to play A by what is precisely the logic of the equiprobability model. The equiprobability theory interpreted thus as an application of level 0 theory is a bounded rationality theory par excellence. A player models her coplayer as bounded—ultra-bounded, because entirely devoid of strategic reasoning; and herself thinks that half the population plays B, a grotesque belief considering that virtually 100 per cent play A. Other empirical evidence also goes strongly against this theory. Stahl and Wilson (1994) found experimentally that over a range of ten games the fraction of subjects who were level 0 was insignificant and the fraction who were level 1 was 24 per cent.[8]

7. Salience Theories

7.1 Applying Variable Frame Theory to Hi-Lo

One idea for explaining why people play A—the behavioural fact—is that it is due to the salience of A.[9] Schelling thought that, however it may be that a particular coordination equilibrium's salience makes it get played, in Hi-Lo this mechanism gets (A,A) played because giving the highest prize makes (A,A) salient. Probably the commonest response of game theorists today to the question why people choose A is an equilibrium-selection version of this idea: the equilibrium (A,A) gets selected because it is the salient equilibrium. This was Harsanyi and Selten's opinion. They point out that each player prefers (A,A) to (B,B),[10] and conclude that we should adopt (A,A) as the solution because the players 'should not have any trouble in coordinating their expectations' at (A,A). (Harsanyi and Selten 1988, p. 81). The suggestion of these theorists is, I think, that salience can explain A-choices by itself, that is, without our having to call in other theories of A-choice, such as bounded rationality, respecification, or payoff dominance. In this section I investigate this suggestion.

The only nonhandwaving theory we have of how rational players solve coordination problems is variable frame theory [described in section 4 of the introduction]. Applied to Schelling games, it works by using salience characteristics to turn these into Hi-Lo games. Could variable frame theory perhaps be reapplied to Hi-Lo games to show that salience characteristics of A and B explain A-choices in them? This might give a nonhandwaving theory of A-choice as an effect of framing.

[It turns out that, provided we allow there to be a family of predicates which includes *giving the highest prize* but not *giving the lowest prize*, we can model the asymmetric salience of A and B in variable frame theory, and that for interesting bands of parameter values we get unique variable frame equilibria. If *giving the highest prize* is highly salient, there is a unique equilibrium in which people almost always play A. To be specific, consider a Hi-Lo game with k basic acts, A, B, ... Let the payoffs be a in (A,A), b in (B,B), . . . , with $a > b > \dots$. Let the universal frame be $F = \{F_0, F_1\}$ where F_0 is the generic family {*thing*} and $F_1 =$ {*highest prize*}; for each player, the extension of *highest prize* in the set of act-descriptions is A. Assume that the availabilities of these families are such that $v(F_0) = 1$ and $v(F_1) > 0$. Then there is a unique variable frame equilibrium in which players for whom F_1 comes to mind opt for *choose the highest prize* while other players opt for *pick a thing*. This implies that an arbitrary player plays A with probability $v(F_1) + (1 - v[F_1])/k$.]

In this theory, the mechanism through which being highest-prized has an effect on choice is that it makes A salient—A is chosen in virtue of this salience. (Coordination on it has to be reasonably well paying, but only above average, not necessarily highest.) I shall call this mechanism for generating A-choices the *pure salience mechanism*. The theory has a perhaps surprising implication: if an act giving a lower prize were for some reason or other highly salient, the low-prize act would be chosen rather than A.

[This can be shown in a simple extension of the previous model. Suppose that each player makes his choice between the options A, B, . . . by pressing one of k buttons. The A button is salient through its association with the highest prize, but another button, R, is salient because it is the only red one. Let z be the average of the prizes $a, b, \ldots$ and assume that r, the prize associated with R, is greater than z (which is possible only if $k > 2$). The reason for this assumption will become clear shortly. Let the universal frame be $F = \{F_0, F_1, F_2\}$, where $F_0 = \{thing\}$, $F_1 = \{highest\ prize\}$ and $F_2 = \{red\}$. Assume $v(F_0) = 1$, $v(F_1) > 0$ and $v(F_2) > 0$.

First, consider a player whose frame is $\{F_0\}$. For her, the only act-description is *pick a thing;* thus she picks among the k buttons at random. Next, consider a player whose frame is $\{F_0, F_1\}$, and thus whose decision problem is {*pick a thing, choose the highest prize*}. In the game as he perceives it, *choose the highest prize* is the unique best reply to both of these act-descriptions. Now, consider a player whose frame is $\{F_0, F_2\}$, and thus whose decision problem is {*pick a thing, choose the red*}. In the game as this player perceives it, *choose the red* is the unique best reply to both of the act-descriptions he can recognise. (To reach this conclusion, we need the assumption that $r > z$.)

Finally, consider a player whose frame is $\{F_0, F_1, F_2\}$, and thus whose decision problem is {*pick a thing, choose the highest prize, choose the red*}. As in the case of the frame $\{F_0, F_1\}$, *pick a thing* is strictly dominated by *choose the highest prize.* (Similarly, as in the case of the frame $\{F_0, F_2\}$, *pick a thing* is strictly dominated by *choose the red.*) So the only act-descriptions that can be optimal for this player are *choose the highest prize* and *choose the red*. Recall that players whose frame is $\{F_0, F_2\}$ opt for *choose the red.* Since the unique best reply to *choose the red* is *choose the red,* if the probability that an opponent has the frame $\{F_0, F_2\}$ is sufficiently high, *choose the red* is optimal for the player with the frame $\{F_0, F_1, F_2\}$. More specifically, let v_T ('T' for 'thing') be the probability that a player has the frame $\{F_0\}$; let v_H (for 'highest') be the probability that a player has the frame $\{F_0, F_1\}$; let v_R (for 'red') be the probability that a player has the frame $\{F_0, F_2\}$; and let v_U (for 'universal') be the probability that a player has the frame $\{F_0, F_1, F_2\}$. That is, $v_T = (1 - v[F_1])(1 - v[F_2])$, $v_H = v(F_1)(1 - v[F_2])$, $v_R = v(F_2)(1 - v[F_1])$, and $v_U = v(F_1)v(F_2)$. Then it is straightforward to calculate that *choose the red* is unambiguously optimal for a player with the universal frame if[11]

$$(v_T/k + v_R)/(v_T/k + v_H + v_U) > a/r. \tag{1}$$

That is, if (1) holds, a player with the universal frame maximises her expected payoff by opting for *choose the red,* irrespective of her beliefs about the behaviour of other players with the same frame as herself. If this is the case, the probability that an arbitrary player presses the red button is $v_T/k + v_R + v_U$, while the corresponding probability for the A button is $v_T/k + v_H + v_U$.

This result is only illustrative: it shows sufficient conditions for there to be a unique variable frame equilibrium with a clear prediction of R, assuming a very stripped-down set of possible frames.[12] In order for (1) to hold, the availability of the 'red' family F_2 must be sufficiently greater than that of the 'best payoff' family F_1 to offset the relatively smaller payoff to be had from coordinating on redness rather than on bestness. For example, suppose that $v(F_1) = 1/3$, $v(F_2) = 2/3$, and $k = 3$. Then (1) tells us that, in variable frame equilibrium, if r is greater than $11a/14$, players who recognise both redness and bestness will press the red button. The probability that an arbitrary player presses the red button is then $20/27$.]

There is empirical support for a pure salience mechanism: Guerra and I gave subjects Hi-Los with $k = 4$, prizes of £6, £5, £4 and £3, and the £5 choice made salient [by the presence of the spade at the top right of the display]. Figure 1.11 shows such a task. [Of thirty-two subjects,

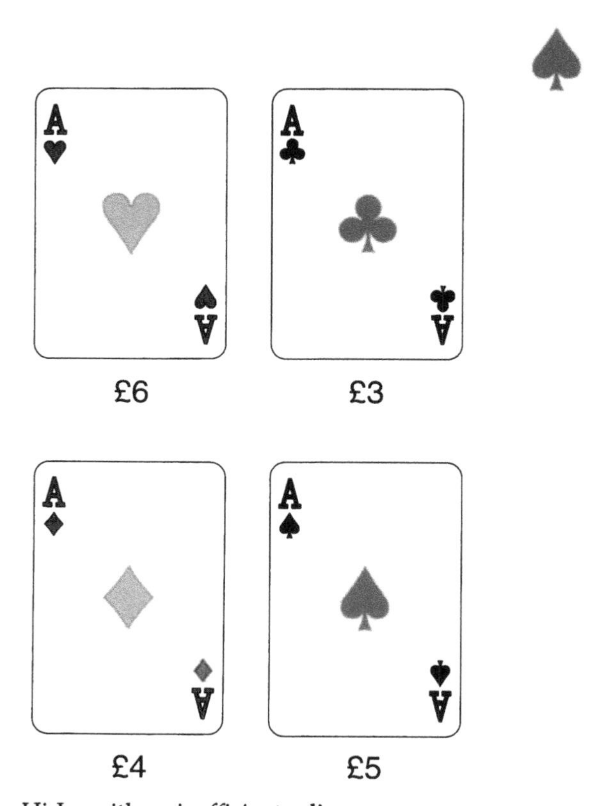

Figure 1.11. Hi-Lo with an inefficient saliency

twenty-three chose the £6 card and eight chose the £5 one.] Making a lower-paying choice more salient significantly dented the hegemony of A, as it should do if all there is going for bestness is its salience.[13]

If the pure salience mechanism were the whole story, then in a context such as the figure 1.11 task, in which another button happens to be sufficiently more salient than the best-paying one, we should feel no qualms about going for it. But this reaction does not match up to our intuitive apperceptions—these seem to be that our attention should be directed towards the best-paying object, even if it's far from the most salient, *because it is the best-paying one*.

Any mechanism in which the best-paying properties have a positive effect on A-choices independently of the salience of A I shall call a *pure bestness mechanism*.

The pure salience mechanism and the pure bestness mechanism are quite different. They are different even if, as usually happens, the best-paying choice and the most salient choice are one and the same. They differ in that they specify different reasons for pressing A. The distinction exemplifies a general feature of the intensionality of reasons. An action may have properties π and π', but if I choose it because it is the π act, then even if I know that the π act is the π' act, it may be false that I choose it because it is the π' act. Indeed I may choose it in spite of its being the π' act.

The fact that the acts supported by pure bestness reasons and salience reasons are highly correlated makes it harder to tell which is operative in naturally-occurring empirical contexts. But because this correlation is contingent and not necessary, it affects not one whit the conceptual distinction.

My own account contains both a pure salience mechanism and a pure bestness mechanism—it says that bestness promotes A-choices both through making A salient and because bestness properties sometimes influence people to choose A independently of salience. Which mechanisms are at work on a given person on a given occasion depends, I shall say, on certain basic features of how she frames the problem. There are two frames. In one salience as such is operative; the judgemental fact is explained by reasons which operate in the other frame.

7.2 *Explaining the Judgemental Fact*

The bounded rationality theories I have described seem to fail to explain the judgemental fact. But the salience theory, though it does not provide an explanation of this fact, points us in the right direction. For it prompts the question: what property of A *makes* it salient? Suppose we found that the salient-making property was a reason-giving property. Two things might follow: (i) we might have an explanation of the judgemental fact, as being due to the salience of a reason-giving property; (ii) salience itself might drop out as an explanation of the behavioural fact, as an inessential by-product of the reason-giving property.

8. A Germinal Theory

8.1 Demands on a Theory of A

There are three requirements for a good theory of why people play A in Hi-Lo: (i) that it imply observed behaviour, that is, the almost universal choice of A in normal circumstances; (ii) that it do so intelligibly to us, which (to the extent that A intuitively and stably seems to us the only rational thing to do) involves displaying A as uniquely rational—that is, giving principles of rationality which are themselves persuasive, and showing they dictate doing A; and (iii) that it be part of a unified theory of a wide range of problems, not just Hi-Lo—for example, all problems of cooperation.

Of the theories of play in Hi-Lo I have been describing, some address only 'What draws people to A?' and so try to satisfy (i) but don't try to satisfy (ii); others concern the question 'What seems right about A?' but in the form 'What seems right about A to players who may be less than fully rational or think their coplayers are?' The judgemental fact to be explained under (ii), however, is not this but that A stably seems right when there is common knowledge of rationality.

Since we have seen that the principles of standard decision theory and game theory fail to predict A, we are driven to conclude either that the apparent rationality of A is an illusion, or else that the rationality of A derives from principles of rationality that are not included in decision theory. This conclusion means that, to meet (ii), a revisionary theory is required. The new principles in such a theory might conflict with established ones, but they need not—they might constitute a revision only in the sense of an extension. Should we aim at this, or should we judge rather that we are under an illusion? Consider for a moment the illusion theory; it proposes that thinking that one ought to play A is a mistake: that all one is entitled to think is that A is rationally permitted—not forbidden, but not positively indicated either. I stick my neck out and posit that A-choosing is rational. Since I reject the illusion theory, this means I must look for a revisionary one. Playing A is rational, I claim, in virtue of considerations and principles not yet identified (the *Revisionary Conjecture*).

8.2 A Conjecture

My own proposed solution of the Hi-Lo paradox is an application of a revisionary theory of how people play games (the 'variable agency' theory) which I will describe in chapter 4. At this point I will describe some broad features of my explanatory strategy, and some reasons why I think we are pushed towards them by various pieces of evidence, including the failures of the previous assaults on the Hi-Lo problem I've been recounting. I will then point out some challenges a theory of the kind I propose will have to meet.

I suggest that (*Conjecture 1*) a typical person facing any game may have an inclination to reason about what to do in it in a certain way, which for the moment I will label reasoning in *mode-P*. I suggest also that (*Conjecture 2*) in the game of Hi-Lo this inclination is very strong. Mode-P reasoning has two features:

F1. The player ranks all act-profiles, using a Paretian criterion.
F2. She takes herself to have a reason to enact her component in the highest-ranked act-profile.

Feature 1 makes mode-P reasoning have something in common with a familiar sort of reasoning about games, the reasoning we engage in when we discuss the 'collectively rational' outcome. A theorist doing this also ranks all act-profiles by assessing the payoff profiles they generate, and on most accounts she does this in accordance with some form of the Pareto principle. We'll see in due course the reason for this similarity.[14] I shall label F1 *collective profile ranking*.

Feature 2 says that if Player 1, say, ranks (x,y) top, she thinks she should do x. I label this feature of mode-P reasoning *projection*, because in mathematical parlance the player's choice is the projection of the whole n-tuple onto the dimension of her act.

In ordinary, best-reply reasoning, a player ranks outcomes according to her own component of the payoff-profile of each outcome. In F1, saying that a player uses a Paretian criterion means that if profile p is Pareto-superior to profile p' then she ranks p higher than p'. Because Hi-Lo is a coordination game, in Hi-Lo this results in exactly the same ranking of profiles as ranking them by her own component. But this is not true in some other games, and in these F1 implies a change in the player's ranking; that is, in these cases mode-P reasoning implies a payoff transformation. For example, the utilitarian function, [which sums the players' payoffs], is one way of meeting the Paretian criterion, and in the Prisoner's Dilemma it gives the ranking $(C,C) \succ (C,D) \sim (D,C) \succ (D,D)$, while the own-component ranking is $(D,C) \succ (C,C) \succ (D,D) \succ (C,D)$.

In the case of Hi-Lo, mode-P deliberation yields A as its unique conclusion. The effect of the Paretian ranking requirement in F1 is that (A,A) is the top element in the player's profile ranking, and F2 then implies that she chooses A. So the first piece of evidence that is explained by this theory is the behavioural fact about Hi-Lo, the fact that there is a very strong tendency to play A. In games in general, the requirement means that an agent in mode-P always selects the unique Pareto-optimal profile when there is one, as there is in all common-interest games. In other cooperation games, such as Prisoner's Dilemma, it restricts but does not fully determine the profile choice; more needs to be said about the player's profile ranking before we know which profile out of (C,C), (C,D) and (D,C) is chosen.

In mode-P a player, Player 1, undergoes not only, in general, a payoff transformation but also a 'reasoning transformation'. F2 implies that she abandons the usual way of reasoning to a best act, best-reply reasoning. Instead, she first selects an *act-profile*, then selects as her act her component in it. This procedure has a quite distinct logical form. Suppose for simplicity that Player 1 has a definite belief about Player 2's act (as distinct from a probabilistic one)—say, that Player 2 will do y'. Let U_1 be a payoff function representing Player 1's ranking in mode-P. Best-reply reasoning has the form: choose the act x that maximizes $u_1(x,y')$. [MB uses U_i to denote player i's ranking of act-profiles and u_i to denote player i's individual payoff.] Mode-P reasoning has the form: choose the act x such that (x,y) maximizes $U_1(x,y)$. In Hi-Lo, best-reply reasoning produces A if y′ = A and B if y′ = B; mode-P reasoning produces A.

8.3 Motivation: Game-Theoretic Indeterminacy

The basic problem we confront is how to explain the apparent determinacy of rational choice in Hi-Lo. A theory should reflect intuitions of rationality by revealing a rational basis for it. If these intuitions are determinate then so should the theory be. Note that this meta-theoretical principle is not the same as the one proposed by Harsanyi and Selten, namely, that game theory should be determinate, full stop: 'Clearly a theory telling us no more than that the outcome can be any one of these equilibrium points will not give us much useful information. We need a theory selecting one equilibrium point as the solution of the game' (Harsanyi and Selten 1988, p. 13).

Contemporary game theory apparently fails because there are two equilibria rather than one. And in versions of it that don't contain the Nash equilibrium principle the failure is worse—all choices are, for example, rationalizable.[15] This is a simply a case of a general property of standard game theory: it is generally underdetermined. That may be fine—we should not prejudge the determinacy of reason. For example, if there are Buridanic cases, [where the agent is unable to identify a strictly best option], then in them our basic rational theory *should* be indeterminate.[16] But in Hi-Lo it's not fine, but monstrous, since the choice of A is so obviously right.

The ultimate source of indeterminacy is, it seems clear, the rational agent's attempt to achieve a best act-profile by choosing only one component of the profile. The agent is modeled as deliberating over—seeking the optimum value of—what she herself controls—her own act. So she must take what she does not control, but is controlled by the other, as parametric, and make a hypothesis about the value of this parameter. But there is nothing that ties down her hypothesis except what she can learn from thinking about the other doing just the same sort of thing. This leads to a notorious circle.

Standard game theory gets rid of much of this indeterminacy by adding the assumption of Nash equilibrium. This is essentially the assumption that each player guesses right. I am not in favour of this assumption for the purposes of rational game theory, but we needn't argue about this here; Nash equilibrium fails to confer determinacy anyway, so there is no need to dispute the principle in order to show that standard theory fails for Hi-Lo.

There are two possible ways out. One, pursued above, is to link the player's own act, in the player's mind, to the whole profile in such a way as to give her reason to think she has indirect control over the whole profile. The classic expression of this is evidentialism, which, however, involves fallacy and so provides at best a boundedly rational explanation of A-choices.

The other, which I pursue, is to consider the deliberator as directly choosing a whole profile. This does not mean that she can implement her choice! That would be at least as magical as evidentialism. But on the other hand she is free to ask herself what would be the best profile—just as the social choice or collective rationality theorist does. And what is also true is that the agents can, *between them*, implement their profile choices—they can if all make the same one and if all implement their components. This process may remind the reader of processes that groups of individuals sometimes pursue in in real time. I shall return to this empirical observation.

This is the broad motivation for mode-P reasoning—for the choice of an act by profile evaluation, selection and projection. I note that for the theory to yield profile choices that can all be implemented does not require that players use Paretian rankings, but only that all players' rankings should have a common top element. But the Paretian feature is one way of yielding such agreement in a very broad class of games (including all common-interest games), and there are other motivations for it to which I shall come in a moment.

We can see in another way why F2—profile selection and projection—is needed, that is, why F1—Paretian profile evaluation—would not be enough by itself. In the case of Hi-Lo, F1 makes no difference to the agent's ranking of profiles; so if a player reasons in the usual, best-reply way, we have got nowhere. Putting it otherwise: to explain the facts about Hi-Lo, a traditional payoff-transformation theory which turns (A,A) into the best profile for an individual player is useless, because (A,A) is the best profile for her anyway.

By the same token, in such cases as Hi-Lo one could for certain purposes do without F1—Paretian profile evaluation—and stay with the agents' original evaluations of profiles. One could do without it inasmuch as F2 would still give the same answer, A. However, we are looking for a plausible theory, not an arbitrary device for predicting A. And it turns out that the theory which makes sense of F2 also makes sense of evaluating profiles in a way that reflects their social virtues, such as Pareto optimality, that is, also makes sense of F1. An additional reason for F1 is that, as we have seen above, in other games F2 by itself would *not* give the same answer. (For example, in the Prisoner's Dilemma each would choose D.)

8.4 The Paretian Ranking Requirement

Why should it be assumed that, even if agents do rank profiles, their ranking is Paretian? Suppose we ask, 'What is it about A that seems to make it the obviously rational choice?' When we consult our intuitions, or examine the responses of subjects to this question, the answers seem to revolve round the property of A that the outcome (A,A) is *best for both* players. Another way of saying this is that (A,A) is the unique Pareto-optimal profile.[17]

The idea that the unique Pareto-optimality, or best-for-bothness, of (A,A) enters into the player's choice of A is also found among the *obiter dicta* of game theorists. Two of the most explicit (as on much else) are Harsanyi and Selten. After they have filtered out nonequilibria, they are still left with two profiles, and ask what reason we could have for predicting (A,A) rather than (B,B). Their answer is that (A,A) is salient in virtue of its Pareto-superiority or 'payoff dominance'. They do not address the question of why payoff dominance should be a salient property. One possible answer is that it is a salient reason-giving property of profiles. F2 is an expression of this interpretation.

To the extent that payoff dominance is reason-giving, the fact of its salience drops out of the theory. It need not do so completely, however, for reasons may be more or less powerful. If the reason payoff dominance gives for choosing A is weak, it may be that the tendency to choose A is boosted by a salience mechanism in which payoff dominance operates as a cause of A-choices by a quite different route. This is just what happens in the theory of this book.

In this theory, certain factors in a situation of interaction make it more likely that an agent will reason in a profile-based way *and* that, if she does, she will order profiles in a Pareto-respecting way. These features include payoff features. Among these payoff features are a common interest, but also harmony of preferences. Hence when preferences are completely shared, as in Hi-Lo, the payoff factors have maximal strength. This part of the theory also has some initial support both from behavioural evidence and from intuitions about rationality. Observed rates of A play are higher than rates of S play in typical Stag Hunts, which in turn are higher than rates of C play in typical Prisoner's Dilemmas.[18] And it seems to be part of our intuition about Hi-Lo that A is all the more compelling because not only is (A,A) best for both of us, but we share exactly the same ordering of all outcomes. Our pro-A intuitions are stronger than our pro-S intuition in Stag Hunt or our pro-C intuitions in Prisoner's Dilemma. Unlike in games like Stag Hunt, where there is conflict of interest over paired comparisons other than those involving the best point, in Hi-Lo there is unopposed potential for mutual gain. The only thing that stands in the way of our realizing it by playing in (A,A) is lack of means of overt communication of the contents of our deliberations; there are no 'real forces'.

All the past theories that attempt to rationalize A-choices treat them, in the mind of the agent, as a way of bringing about (A,A) rather than some other profile. But the effect of the Paretian requirement in F1 is just this—it makes the target profile be (A,A) rather than some other. So, it may be asked, is there any point in bringing in the Paretian requirement? The answer is yes, because of the nonextensionality of reasons and the need to model agents' operative reasons. In these theories, the reason why the pair (A,A) is an aim for Player 1 is the property of (A,A) that it maximizes; it isn't its Paretian property (that it is uniquely Pareto-optimal). It might be that a player chooses A for the reason that (A,A) has one of these properties without choosing it for the reason that it has the other. This is true even when, as here, she knows that whatever has one has the other.[19]

What sort of a reason for choosing the profile (A,A) does its Pareto-optimality give? Why should an agent who ranks profiles rank them in a Pareto-respecting way? The very name suggests that such an agent is swayed by considerations of collective or systemic rationality with respect to the collectivity consisting of both (in other cases, all of the) players. It will be part of my thesis that she is.

8.5 Challenges the Theory Faces

A theory which contains 'mode-P reasoning' faces many serious challenges. First, closure: as it stands it appeals exogenously to an inclination to enter mode-P; we need a proper theory of the factors that determine the inclination to reason in this way. Second, collective profile ranking: it remains to be explained why, if and when people start to reason in mode-P, they should be swayed by considerations typical of collective rationality. Third, projection: even granted that an agent should decide that a certain profile p should be realized, why does this give her a reason to do her part in p; given that she favours (A,A), why does this mean she should do A? The question immediately raises a doubt about the whole idea of mode-P reasoning. It's no good, usually, playing your part unless others will too; it will not by itself achieve the desired profile, and it may well be counterproductive in terms of your ranking of profiles. For example, if P2 will in fact choose B, choosing A produces a worse profile than choosing B does if your ranking is the shared ranking given by both u_1 and u_2. Deliberating in mode-P can be futile or worse than standard-mode deliberation if other players are not deliberating likewise. Thus any inclination to do so needs to be tempered by suitable caution. How to include such tempering with caution is a fourth challenge for the theory.

*APPENDIX
Payoffs in Calling a Catch

In the Calling a Catch game, a *state of the world* can be defined by three components: whether F1 really is better-placed (denoted by 1) or F2 is better-placed (2), whether F1 thinks he is better-placed (*b*) or worse-placed (*w*), and whether F2 thinks he is better-placed (*b*) or worse-placed (*w*). This gives eight states, for each of which the prior probability can be calculated straightforwardly. For example, (1,b,b) is the state of the world in which F1 really is better-placed, but each player thinks that he himself is better-placed; the probability of this state is $(1/2) \times (2/3) \times (1/3) = 2/18$. Each player chooses one of the strategies B (go for the catch if and only if you think you are better-placed), W (go for the catch if and only if you think you are worse-placed), or—an option that is allowed only in the second

TABLE 1.1
Payoffs for calling a catch

		Payoff if strategy profile is					
State of the world	Probability	BB	BW	WW	BC	WC	CC
(1,*b*,*b*)	2/18	−5	20	−2	20	−2	10
(1,*b*,*w*)	4/18	20	−5	15	20	−2	10
(1,*w*,*b*)	1/18	15	−2	20	−2	20	10
(1,*w*,*w*)	2/18	−2	15	−5	−2	20	10
(2,*b*,*b*)	2/18	−5	15	−2	−5	10	10
(2,*b*,*w*)	1/18	15	−5	20	−5	10	10
(2,*w*,*b*)	4/18	20	−2	15	10	−5	10
(2,*w*,*w*)	2/18	−2	20	−5	10	−5	10
Expected payoff		9	5.8	7.3	8.8	2.7	10

version of the game—C (obey the captain's call). Since the positions of the players are symmetrical, it is sufficient to consider only six strategy profiles: BB (both players choose B), BW (F1 chooses B, F2 chooses W), WW, BC, WC and CC. Given the state of the world and the strategy profile, each player's behaviour is determined. If one and only one player goes for the catch, the payoff for the team is 20 if this player really is better-placed and is acting on B or W; it is 15 if this player really is worse-placed and is acting on B or W; and it is 10 if this player really is better-placed and is acting on C. (A player who really is worse-placed is never called by the captain.) If neither player goes for the catch, the payoff is −2. If both go for the catch, the payoff is −5. Table 1.1 shows, for each state of the world, its probability and the payoff, given each strategy profile. The expected payoff for each profile is shown in the final row.

Notes

1. Sometimes a safe single is not taken because this makes the other batsman the striker, and he may be weaker than the current striker. But this means the payoffs are not as shown.

2. We will come back to possible explanations of why it was 91 per cent rather than 100 per cent. Earlier tasks were matching games in which all prizes

were equal, but one was made salient. In the figure 1.8 task, the £5 card is salient in the spatial configuration. As we shall see in chapter 4, Bacharach and Guerra found that salience can override prize size.

3*. MB was planning to add here some direct evidence that people think A obviously rational.

4. The main tradition in game theory also holds that this is the only kind of reason that can rationalize an action, but agreeing that it is one sort that can does not commit one to this view.

5. Figure 1.a illustrates this case: T is Player 1's Stackelberg act, L is Player 2's, but (B, L) and (T,R) are the only equilibria.

		Player 2	
		L	*R*
Player 1	*T*	1, 1	3, 2
	B	2, 3	1, 1

Figure 1.a. A Stackelberg-nonsoluble game

6. In fact in every common interest game the Stackelberg acts are in equilibrium, and this is the common-interest equilibrium, so the conjecture predicts S in Stag Hunt, however risky S may be. The conjecture is therefore too strong as it stands; however, it remains possible that Stackelberg reasoning occurs, but can be overridden by other considerations.

7. For example, in the Stag Hunt of figure 1.10, (R,R) is risk dominant, because adhering to (S,S) against a coplayer equally likely to be an adherent of (S,S) or (R,R) implies playing S against equally probable S and R, and similarly for adhering to (R,R).

8. It might be possible to exploit the equiprobability idea in a way that does not run into these difficulties by thinking of equiprobability as the initial belief state of a player in an iterative process. Harsanyi (1975) suggests that players generally come to their final beliefs about what each other will do through such a process, called the tracing process. Each begins with a prior distribution over the other's acts, then computes her best reply, then her coplayer's best reply y to this; she then bends her prior somewhat towards y, and repeats the whole process. At each stage the reasoning of a player is level 2, which is a modest but decent level, modal in the distribution found by Stahl and Wilson (1994); but the players also have some inconsistent beliefs: when Player 1's probabilities for Player 2's act are π_t, she also thinks Player 2 has beliefs on which π_t is not optimal, and that Player 2 is a rational optimizer (Bjerring 1977). In other respects the process is too demanding.

9*. In this subsection, the material enclosed by square brackets revises MB's exposition so as to maintain consistency with the treatment of variable frame theory in the introduction. These changes are matters of notation and theoretical detail, not of substance.

10. Mutatis mutandi—their example is not Hi-Lo but a Stag Hunt.

11*. Consider a player who has the universal frame. With probability v_T, her opponent has the frame $\{F_0\}$ and opts for *pick a thing*. With probability v_H, her opponent has the frame $\{F_0, F_1\}$ and opts for *choose the highest*. With probability v_R, her opponent has the frame $\{F_0, F_2\}$ and opts for *choose the red*. To make it as difficult as possible to show that *choose the red* is optimal, assume that if the opponent has the universal frame he opts for *choose the highest*. Then the expected payoff from *choose the red* is $v_T\, r/k + v_R\, r$, while that from *choose the highest* is $v_T\, a/k + (v_H + v_U)a$. The former is higher if (1) holds.

12*. Because this result is concerned only with sufficient conditions for the optimality of *choose the red*, it does not require the principle of payoff dominance to be used. By presenting his analysis in this form, Bacharach avoids having to resolve the issue of how payoff dominance should be understood when the availability of a family of predicates is less than 1. This issue is discussed in section 4 of the introduction, in relation to the game of Large and Small Cubes.

13*. This experiment is described more fully in section 8 of chapter 4.

14. We also find a hint of feature 1 in the equilibrium selection literature, in which it sometimes seems that it is the players, rather than the theorist, who are supposed to select among equilibria—in which case they are also selecting among profiles. But this is nowhere spelled out, and is not the standard interpretation of 'equilibrium selection'. Harsanyi and Selten themselves occasionally veer towards the players-as-selectors interpretation. In discussing payoff dominance they write: 'Clearly, among the three equilibrium points of the game, U_1U_2 is the most attractive one for both players. This suggests that they should not have any trouble coordinating their expectations at the commonly preferred equilibrium point U_1U_2' (Harsanyi and Selten 1988, pp. 80–81). Here Harsanyi and Selten seem to come near to proposing a profile-selection theory, that is, a theory in which players vet and evaluate alternative profiles, and this evaluation governs their decisions.

15*. The concept of rationalizability was introduced to game theory by Bernheim (1984) and Pearce (1984). In a game for two players P1 and P2, a strategy r is rationalizable for P1 if all of the following conditions are satisfied: (i) r maximises P1's expected payoff, given *some* probability distribution over P2's strategies; (ii) every strategy of P2's that is assigned a strictly positive probability in (i) maximises P2's expected payoff, given *some* probability distribution over P1's strategies; (iii) every strategy of P1's that is assigned a strictly positive probability in (ii) maximises P2's expected payoff, given *some* probability distribution over P1's strategies; and so on. Rationalizability differs from Nash equilibrium in not requiring the two players' beliefs to be consistent with each other. In Hi-Lo, for example, it is rationalizable for P1 to choose A (in the belief that P2 will very probably choose A) while P2 chooses B (in the belief that P1 will very probably choose B).

16. True, even for Buridan cases we need, as agents, higher-order reasons, which tell us what to do when lower-order reasons fail to pinpoint. But this does not mean that theory should at all costs introduce further, tie-breaking reasons into the basic theory. If we should, then they are not Buridan cases, *contra hypotesi*—and there are never any Buridan cases.

17. The profile (x^*,y^*) is uniquely Pareto-optimal if and only if for all $(x,y) \neq (x^*,y^*)$, $x^* > x$ and $y^* > y$.

18*. MB planned to insert a footnote documenting this claim.

19. Indeed, this is what classical decision theory obliges us to say. Imagine a situation in which I know that by choosing x I bring it about that my coplayer does (for instance, I know he copies me out of sycophancy), and suppose that (A,A) maximizes both my and my mimic's payoff, both u_1 and u_2. According to decision theory the only reason I can have for action is to bring about consequences I prefer, and u_1 fully represents my preferences about consequences. If I happen to be a Player 2-sympathizer and prefer that she gets what she wants (or a Player 2-antipathizer and prefer that she doesn't), these preferences are already 'in' u_1.

Chapter 2

Groups

THIS CHAPTER concerns the place of groups in our life, and in the way we model our life. It begins with the propensity of humans to think of individuals—both themselves and others—as members of the groups of which they are members: the propensity of people to 'group identify' people. My own theory of successful group activity—of cooperation—will have as a basic building block group identification of oneself—thinking of oneself as a member of a group. This theory focuses on certain production conditions for, and effects of, this phenomenon: that it is a framing phenomenon, with all the characteristic variability and discreteness of framing; that it determines choice; and that it does so by changing the logic by which people reason about what to do. The theory is concerned with intragroup processes rather than intergroup relations, and processes that don't depend essentially on long-term relationships. But in this chapter I will not confine myself to these; I will sketch the landscape of group identity theory, such is the importance of this theory not only for my particular investigation, but for the general understanding of human society.

1. Framing One's Self

In order to explain how someone acts, we have to take account of the representation or model of her situation that she is using as she thinks what to do. This model varies with the cognitive frame in which she does her thinking. Her frame stands to her thoughts as a set of axes does to a graph; it circumscribes the thoughts that are logically possible for her (not ever, but at the time). In a decision problem, everything is up for framing. The preferences on which she acts, her alternatives, and the alternatives of her coplayers, all are. So far from finding herself with given preferences over outcomes, as traditional theory holds self-evident, these preferences depend upon the evaluative concepts that are uppermost in her mind. But nor is this all; also up for framing are her coplayers, and herself.

I shall suggest later that the answers to fundamental questions about coordination and cooperation, including the resolution of the Hi-Lo paradox, lie in the agent's conception not of the objects of choice, nor of the consequences, but of *herself* and of the agents with whom she is interacting. In this chapter I shall consider some of the alternative ways in which an agent can frame the interacting agents. The effects of these alternative framings are profound, for they concern the very logical structure of the problem she faces. They have to do not with the personal characteristics of the agents but with their boundaries. One such agent-boundary issue is whether an agent who is deliberating at a particular time thinks of herself as an agent with a past and future—a person—to do which she must have a frame that includes the notion of person. Another issue concerns the way in which an agent's group memberships enter into her conception of herself.

I will now sketch the varied landscape of the literature of groups. We find groups of many different sizes and kinds, portrayed sometimes as nonthings and sometimes as things, sometimes as good things though more often as bad. Prominent among the psychological features of members of groups, we find the phenomenon, crucial for the argument of this book, of 'group identity'. I next describe the characteristic features of group identity that have been the subject of attention in the past. Of these the most fundamental is that one who group identifies *conceives herself* in a quite different way from one who does not, so that group identification involves a shift of frame.

2. Human Groups

2.1 Entification from the Outside

A *group* is a set of objects that we 'entify'—that is, treat as an entity. In grammar, it can take a singular verb. It is perhaps the most widely applicable word that does the entifying job; there are countless others for particular domains—archipelago, brotherhood, crowd, drove, eleven, firm, galaxy, herd, interest group, . . . , yeomanry, zemstvo. We can even complete our alphabet of group terms with 'xiphias' or 'xoanaon' or 'xyrid' (a swordfish, a kind of carved image, a kind of iris), once we recognize that middle-sized objects such as these are themselves aggregates of smaller particles. This may seem a bizarre way to think about them, but that is just because in cases like xyrids our entification is particularly strong.

Not just any set of items is described as a group, or by another word with the same grouping function (a committee, a xoanaon), so it has been several times asked: what causes a set to be taken as a group? That is, what makes us entify? Some fundamental group-making principles have been proposed. The Gestalt school noted (Wertheimer

1923) that entification in visual perception is involuntary, and proposed that features of sets producing it include 'contiguity', 'common fate' (moving in parallel over time), 'good figure' (forming a recognizable pattern) and 'similarity'. Thus a set of dots on a screen that moves about together maintaining an R shape is seen not *as* that, but *as* an *R*. We so see it whether we like it or not; we are helpless before the combined entifying force of contiguity, common fate and good figure. They make the *R*-ness of the configuration overwhelmingly salient. Other features of objects have also been proposed for perceptual entification, such as closedness and impermeability (Campbell 1958).

'Similarity' is unlike the other perceptual criteria, as it leaves open and unspecified the dimension of similarity. Post-Gestalt psychologists (Campbell 1958; Tajfel 1969; Tversky 1977; Rosch 1978) make similarity the basis of a rather general criterion of grouphood, the *meta-contrast* principle. This principle explains categorization, that is, the cognitive activity of dividing a domain of items into two or more groups (and so simultaneously creating more than one new entity). The principle is: given a set *S* of items, we are more likely to make entities of subsets of *S* the smaller are intrasubset differences relatively to intersubset differences. Here 'difference' is defined in terms of a metric space of saliently different attributes. Categorization is essential for getting through life; but it is also introduces biases into our judgements, in particular 'accentuation effects': once someone has formed a category *S*, she tends to perceive greater similarity between members of *S* and greater differences between members and nonmembers.[1]

If the elements of the set are humans, in addition to the spatio-temporal Gestalt properties—contiguity, common fate and so on—psychological or social analogues of these properties tend to make us see the set as a group. Particularly important are common fate in its everyday sense—all members of the set being bound to experience the same outcome, nice or nasty—and good figure in cases in which the 'pattern' is an organizational structure.[2] Likewise psychological, cultural or social, rather than spatio-temporal, similarities, such as all being Muslims, tend to produce entification of human aggregates. A dimension of similarity that can have no analogue in inanimate objects is that of goals; a common goal is sometimes taken to be a crucial condition for counting as a group.

We have a particularly strong tendency to see an aggregate of individuals as a single entity when the aggregate displays the characteristics of an 'organized system'. By this I mean, roughly, a system whose behaviour can be explained in terms of a goal and a pattern of information processing which yields actions that subserve that goal. Examples of such aggregates are the sets of individuals we call football teams, firms, political parties, and nations. By no means do we see all sets of individuals that we entify as organized systems: we group people into families easily enough, but cases of families that are organized systems are relatively rare, though we all know a few. [MB planned to] argue later that persons themselves are seen as entities because they are organized systems made up of certain more elementary components. The strength of the tendency to entify an organized system is not surprising, since organized systems simultaneously satisfy three of our criteria—common goal, common fate and organizational structure.

What I have been discussing is what makes ordinary people perceive or conceive a set of things, and in particular a set of human agents, as a group. This is of great importance for our investigation, because we will need to discover the circumstances in which a set of ordinary agents is seen as a group by one of the members of the set. But before proceeding to this question, let me mention some of the groups that crop up not in the world of the layperson but in the world of social and affine sciences. We find 'face-to-face groups', clans, tribes, superorganisms, language communities, societies, persons, 'in-groups', 'out-groups', firms, cartels, interest groups, pressure groups, institutions, panels of experts, teams, informal workgroups, and 'multi-agent systems'. Whether or not groups are a fact of life they are certainly a fact of theory.[3] Of the major disciplines concerned with social behaviour, game theory alone, it seems, has eschewed, and got along to date without, group notions. In our attempt to explain the world about us as professional scientists of many disciplines, as much as in our practices as lay scientists, it seems we cannot do without them. My own theory of coordination and cooperation will draw on most of this plethora of group notions and connect several of them.

2.2 *Entification from the Inside*

Consider any collection of people. What does it take for one of them to think of the lot of them as an entity? If so, does this make a difference to anything—to their attitudes, to their behaviour? We shall find that the answer to the second question is that it can make a profound one.

The lifeboat. Eight people are in a lifeboat, the *Cormorant*, on the high seas, strangers to each other before the ship was abandoned and the passengers and crew took to the boats. In the near distance they can see two other lifeboats from their ship, like their own but of different colours. A helicopter pilot, had there been one, would soon entify our eight; no problem, for there are three quite distinct lifeboats, and our eight can be clearly seen in one, though it's hard to tell exactly how many are in it from this distance. He might call them to himself 'the people in the red lifeboat'. For him they would also be the ones that need to be rescued from the red lifeboat, whose common fate is in his hands. In the lifeboat, it is common knowledge that land is to the west, and that the best chance of survival is to row into the sunset with a steadfast stroke. Each rower entifies the eight of them as 'the eight of us in *Cormorant,* who need to row west with all our might'.

So far the only difference this story provides between entification from the outside and the inside is in the clutch of properties that each uses to define a group. In one case it includes being in the red boat and needing rescue, in the other it includes being in the boat called *Cormorant* and needing to row. But there is likely to be a further difference. An insider is likely to undergo a change in the way she conceives or identifies herself. She may 'group-identify'. Group identification may in turn have a marked effect on her behaviour; it may, for example, lead to her contributing to the common cause by rowing with all her might rather than free-riding on the efforts of others. But mere entification from within is far from being a sufficient condition for group identification. If I am Shelley bullied at Eton, I entify the bits and pieces of Eton into an institution but do not self-identify myself as a member of it; if I am a disaffected member of the Labour Party, I entify the party but no longer feel a part of it. In the lifeboat, it is, we feel, that the group has a common goal which can be realized together that somehow lies behind the rowers' identification. We shall find that this is part of the story.

3. Group Identification

3.1 Self-Identity

In the self-categorization theory of Turner et al. (1987) in psychology, saying someone has a certain 'group identity'—say, that of Anarcho-Syndicalist—means that she identifies herself as a member of that group, that is, this is a way in which she conceives herself and distinguishes herself from other people. This theory is a cognitive theory par excellence. In such cases, the person thinks of herself as an Anarcho-Syndicalist, or Tottenham Hotspur supporter, or whatever, and thinks of her properties as that subset of her properties which are characteristic of members of that group. In a polar case, in which a group membership provides the way and not just a way of conceiving herself, it follows that she regards herself as interchangeable with other members of the same group. According to Turner et al., she has a self-representation in which her self is seen as an 'interchangeable exemplar of the category'. She perceives this equivalence even though she knows she is 'numerically' distinct. This depersonalization is held to explain a variety of attitudes and behaviours observed in groups, to which I am about to come.

However, the theory that people identify themselves as members of groups is not wedded to the polar case, and indeed regards self-definition involving multiple group memberships as normal. Moreover, group memberhips are only one way in which people define their 'selves'. Brewer and Gardner (1996) distinguish three types of self-identity: personal, relational and collective (group). A person's *personal* self-concept is her sense of unique identity due to aspects of her representation of herself that differentiate herself from all others; group memberships are only one kind of differentiating aspect. Someone's *relational* identity is a self-conception in terms of relationships with other individuals with whom she interacts.

Nor is it the case that a given person has a single self-concept; self-conception is self-framing, and shares the characteristic instability and context-dependence of all framing. To be precise, the concepts I deploy at a time to represent my self are a subset of my current frame, which is variable. The principle of meta-contrast is said, in the self-categorization theory, to operate within this frame.

I may be the unique person me at one moment, and a department man or Tottenham man or family man the next. Which of my collective personae is activated depends on the current 'accessibility' of the categories to which I belong: department, football club, family (Bruner 1957). Gurin and Markus (1988) find the social categories people use for self-identification satisfy Bruner's hypothesis that the relative accessibility of a category depends upon many things, which include the perceiver's current expectations, tasks and purposes. In human interactions, the accessibility of categories is a special case of the notion of availability of frames at the heart of the variable frame theory of games. The process in which categories are activated is context-dependent and jumpy.

In the lifeboat, a group is created by the mental processes of its occupants. The mental processes in question are each person's identifying herself and the others as the people in the *Cormorant*. What makes these occupants *be* a group is no more than the occurrence of these mental processes. Group self-identification thus has this strangeness: when a number of us begin to think of ourselves as Cormoranters it is not that there already is such a group which we then use as a way of conceiving ourselves. The acts of self-conception are *accompanied by* the coming into existence of the thing we conceive ourselves by. This self-reflexive property is used by Simmel (1910) and Searle (1995) to define a 'social group'. Searle says that a set *S* is a social group *G* if and only if members of *S* sometimes take themselves to be members of *G*. Simmel says: 'the consciousness of constituting with others a unity is the whole unity in question in the societary case' (1910, p. 374).

3.2 *Identities and Goals*

We shall see shortly that a cardinal feature of group identification is to take the goals of the group to be your goals. This must be so in cases in which part of what it is to be an *X* is to have certain goals. You can't be a Tottenham man without wanting Spurs to win or a department man without wanting the welfare of the department. Furthermore, there are effectively no groups with identical goals, so the variability of my group identity implies the variability of my goals. Variable identity is thus one specific way in which the claim of variable frame theory that goals are frame-dependent comes to be true. If I see things as a department man at one time and as a family man at another, then at different times I have different goals or, at least, different occurrent goals. There are moments at which my colleagues and I, the whole ragbag of us, see the world through departmental eyes, and moments at which the rags occupy other bags.[4]

These are cases in which my self-defining group is, or is seen as, an entity out there. In this case it is often seen as having goals (a football club), or as being subject to a common fate (an oppressed minority). Things are different in cases of 'instant groups'. Here some aggregate of people are placed in a situation in which they see themselves as a group which was not there before but is quite novel. There is no necessary reason why novel groups of these kinds should be endowed with goals or fate by the entifier. It seems, however, that an important source of innovative entification is homogeneity of existing goals. This homogeneity follows from the general principle of similarity, but there is more to it than that. A prominent hypothesis in the literature is that having common interests promotes group identification. For example, five people stuck in a lift see themselves in this way. 'Five people in an elevator are an aggregation, not a group. . . . The same five people stuck in the elevator between floors are a group . . . , the group being shared fate' (Caporael 1995). A sort of converse of this seems very plausible: the greater the disparity of existing interests in an aggregate of persons, relevant to current powers—the more competitive is the game the people find themselves in—the less likely is it that they innovatively identify as a group.

3.3 *Classical Causes and Effects*

It is an established theory in social psychology (the broad literature is reviewed by, for example, Oakes, Haslan and Turner 1994; Brewer and Miller 1996) that there are certain circumstances which tend to produce 'group identification', and group identity in turn produces certain judgements, attitudes and behaviour. 'Group identity' is not directly observed; but these claims mean that it is a theoretical term in a theory which can be judged empirically as a whole, by its ability to explain and predict these effects in the presence of these influences.

Among the favouring influences are being members of the same pre-existing social group; belonging to an ad hoc category (Tajfel 1970); exposure to the pronouns 'we', 'our', and so on (Perdue et al. 1990); having 'common interests'; being subject to a 'common fate' (Rabbie and Horwitz 1969); shared experience (Prentice and Miller 1992); face-to-face contact (Dawes, van de Kragt and Orbell 1988); and 'interdependence' (Sherif et al. 1961), that is, having common interests that can only be achieved together. The effects of group identity include judging oneself to be more similar to other group members and dissimilar to nonmembers than one really is (accentuation), seeking unanimity at all costs (groupthink), favouring members against outsiders in bestowing benefits and in judging worth (in-group favouritism), using first-person plural pronouns,[5] depersonalization,[6] emotional contagion,[7] adopting the group's norms, being motivated by the group's goals (payoff transformation), and cooperating with fellow members. To these I will later propose to add another: agency transformation. Figure 2.1 shows the overall structure of this theory. Although there is wide agreement that there is such a psychological condition as having a group identity, and that it is characterized by some of these causes and some of these manifestations, there are unresolved disagreements about which are the basic ones.

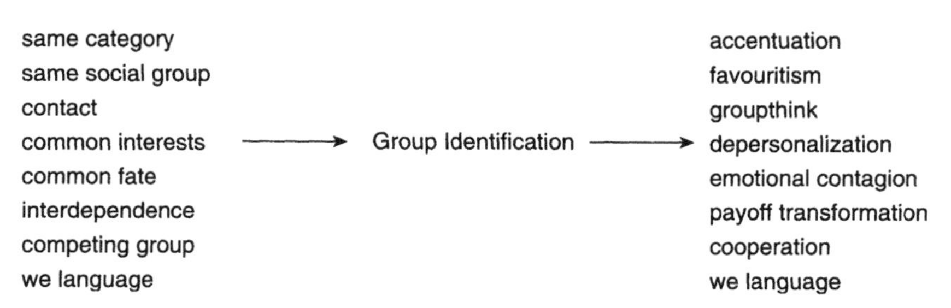

Figure 2.1. Sources and effects of group identification

In 1970 Tajfel showed that being co-classified with others leads a person to discriminate in their favour even if the basis of classification is arbitrary and, in common sense, irrelevant to the decision. In Tajfel's experiment twelve-year-old Bristol schoolboys were told they had either over- or underestimated the number of dots in displays flashed on a screen, then were invited to divide a sum of money between two other subjects, identified only as over- or underestimators: twenty-six out of thirty-two subjects gave more to same-group than to other-group targets. Other experiments showed that this arbitrary categorization treatment, known as the 'minimal group paradigm', also produced other characteristic effects of group identification, including evaluation bias (Doise et al. 1972), and that this can also be produced by exposure to first-person plural pronouns (Perdue et al. 1990).

Tajfel's work challenged two then-prevailing views about the basis for discrimination in favour of one's own group against nonmembers. The accepted wisdom was that, first, group identity and its effects developed over time with repeated interactions among the agents involved. Second, it was thought that the basis for the psychological formation of the group was the interdependence of individual interests. The early theories thus emphasized personal contact over time among members of the group, a history of contact with another group, a common goal for group members, and conflicting goals between the groups. In the minimal group paradigm all this is missing: the tasks are one-shot and the subject's own material payoff is unaffected by her decision. Groups can apparently be instantly formed through a purely cognitive, non–goal-involving mechanism.

The diachronic, interest-based account of intergroup relations had been stimulated by another experimental paradigm, which originated in the well-known Robbers Cave experiment of Sherif et al. (1961). In this experiment twenty-two white middle-class eleven-year old boys, on arriving at a summer camp, were randomly divided into two groups. For a week each group was given cooperative tasks and the two groups were kept at a distance. Each group was left to its own devices, and each developed a group identity and learned how to cooperate. I shall return in a moment to these cooperative practices. At the end of the week the two groups of boys were brought together in a competitive game; they at once manifested great hostility towards each other, from that day on fighting, name-calling, and refusing to go on daytrips if the other group was to be present.

There is, then, disagreement between cognitivists about group identity who, like Tajfel and Turner, hold that mere categorization is enough to produce the characteristic effects of group identity, and interdependence theorists, like Rabbie, Schot and Visser (1989), who hold that common perceived individual interests are essential in the explanation of these effects. This division is correlated with another: cognitivists hold that group phenomena are *sui generis*, while interdependence theorists are often individualists who hold that group identity and its manifestations are entirely reducible to individual interests and sentiments (Lott and Lott 1965). But this school of thought, though denying a privileged theoretical status to groups, does not deconstruct group identification; its members think that the group interest to which identifiers subscribe agrees with the individual interests of the identifiers, and that this agreement is part of the key to the identification phenomenon. Contrast Hardin (1995), who argues, for macro-social groups such as ethnic and religious groups, that people who join them typically have private motives (of which they need not be aware) at variance with the group's public goal, and that these private motives lead them to join because membership affords them opportunities to fulfil them.

Both the Sherif and the Tajfel studies, and most other experimental work on group identity, have focussed heavily on so-called intergroup situations: situations in which persons who identify with one group form attitudes or act in an interaction with members of another perceived group, the out-group. This focus is not just a historical accident: some believe that there is no in-group without an out-group. 'We are what we are because *they* are not what we are' (Tajfel and Forgas 1981, p. 124).

This intergroup perspective on the group is one of the things that has given groups a bad name. Among the best-confirmed characteristics of in-groups are discriminations against outsiders: bad feeling towards others, issuing in overt hostility to others for no better reason that that they belong to out-groups, and negative biases in judgements of others, that is, valuations that are systematically worse than the objective evidence warrants. We have daily confirmation of these detestable attitudes and practices.

But the intergroup perspective may itself be criticised, as a partial frame of the theorist, which distorts the truth by masking off important properties. There is evidence that the thesis that group identification depends on there being an out-group is much too strong, and that the presence of a particular distinct out-group only enhances the process or effects of group identification. Brewer (1991) thinks that, in particular, 'the perception of common fate and shared distinctiveness may be achievable without reference to specific outgroups, although the presence of outgroups may influence the nature or intensity of affect or emotions attached to ingroup membership'. If this is so, two things follow. First, deploring our disposition to sink our individual identities in groups may itself involve a biased judgement, since it is based on concentrating excessively on features of group identification which depend on there being an out-group. If these are typically deplorable, like negative stereotyping and discrimination, or some out-group–independent features are admirable, group identification will be judged too harshly. And indeed, one claimed good effect of group identity, cooperation, appears to be possible without an out-group. Most of the experiments which have shown that group identification enhances cooperativeness have none. Second, it should be possible in principle to bring about a sense of superordinate group identity between members of distinct existing groups. And indeed several experiments, beginning with Sherif's, have shown that even where there are competing groups, introducing a shared goal can produce cooperation in achieving it.

3.4 Empirical Research on Group Identity and Cooperation

It is claimed on both sides of these debates that one characteristic effect of group identification is more cooperative behaviour within the group. Most of the experimental work on this question has been essentially qualitative, seeking to establish the presence of effects rather than to measure responses to measured stimuli. The early work was in the field, but since Tajfel (1970) showed how to manipulate instant group identity, there has been a steady stream of laboratory tests.

In the 1930s the most influential research on cooperation in groups was by organizational behaviour theorists interested in cooperation in the workplace. These investigators did not think of group identity as a mediating state, caused by various factors and giving rise to various effects, as in figure 2.1; rather, they thought of it as being one of the features that emerged in a group during a diachronic process of 'group formation', together with other features like cooperative behaviour and hostility to out-groups. Mayo and his associates' observations and interviews at the Hawthorne plant of Western Electric (1933) revealed that within the formal working structure set up by management had developed cliques—informal, gang-like groups—which had their own objectives (such as restricting output to maintain piece-rates and minimizing effort subject to satisfying management). These informal groups had created an internal organization consisting of an authority structure and a system of norms and sanctions, and this functioned successfully to achieve their common goals (Roethlisberger and Dickson 1939). Each had a leader and 'side-kicks'. The norms included not being a 'rate-buster' (working too much), a 'chiseler' (working too little) or a squealer. Their sanctions included 'binging' (tapping on the upper arm), ridicule and social exclusion. The cliques showed high 'cohesion', the desire to stay in the group born of a 'desire to belong' and of the attraction of the other members. We may say that what developed was a group which displayed group identity born of a common goal and a way of cooperating to achieve this goal, namely a certain internal organization.

Sherif et al. (1961) observed much the same process of group formation in their groups of summer-camping schoolboys, but found that in this case each group pursued the goal assigned to it by the camp organizers. The organizers nudged each group towards identification by inviting them to name themselves; they chose the names Rattlers and Eagles. Over the week that this phase of the experiment lasted, the schoolboys developed positive attitudes towards their own groups. Each group cooperated internally through a self-generated organization involving leadership, norms and sanctions; different leaders emerged for different cooperative activities (cooking, fishing), which they coordinated; norms included favouring certain songs and jokes;[8] sanctions, such as making deviants remove stones from the swiming pool, fitted the circumstances.

These studies emphasize common aims, endogenous or exogenous, as forces making for group formation and cohesion. But aims are never fully common, even in a lifeboat, much less in a large enduring formal organization. Identification may not be complete, or it may be to begin with but then wither away. The freedom fighter may be concerned one day for his own skin. Individual members may and do try to move or use the group for private ends: they may have private and even hidden agendas (Jones 1973). More generally, we may see the behaviour of an enduring group as the joint product of two sets of influences, from the group on its members and from the members on the group, both acting through a variety of channels (Van Winden 1983). A trade union might have average wage and total employment as objectives; these goals might lead employed members to care about the employment of comembers who might be out of work and so put less weight on their own wages, either from enhanced sympathy for fellow members or by internalization of group norms. In the other direction, if the unemployed drop out of the union as they become outsiders in the labour market, the union's objective function may gradually put more weight on average wage.

The spontaneous groups in the bank wiring room at Hawthorne had developed gradually, over months; at Robbers Cave, the process of group self-formation was observed for a week. But it is perfectly possible for a group to self-form in an hour or so or even in a few minutes or seconds. The emotion of shame at observed misconduct, often thought to act to cement groups together, is the psychological product of changes in the autonomic nervous system and hormonal processes which can start within two hours of the birth of a group (Campbell 1994). In the experiments of the Dawes group, which I discuss in the next paragraph, ten minutes of talk were allowed before the main task was administered. And if three able-bodied passers-by, strangers to each other, observe an elderly lady whose car is stuck in the middle of traffic, they may spontaneously and without exchanging words form a group in a few seconds whose founding mission is to give her a push start.

Both these research projects [the Hawthorne plant and the Robbers Cave] were field experiments, and the limited control that is inevitable in such studies makes it hard to draw hard conclusions about causal relationships among the input factors, group identification and behaviour. But Tajfel's 1969 categorization experiment initiated a phase of laboratory experiments, and a now substantial series of these have zeroed in on some of the detail of these relationships. In a number of them the behaviour is choice behaviour in games with scope for cooperation. Collectively they support the figure 2.1 schema by showing that one or another of the figure 2.1 influences enhances rates of cooperation in these games. The stimuli used to prime group identity have included plural framing (Brewer and Kramer 1986; Cookson 2000),[9] joint membership of an existing social group (Wilson and Katayani 1968; Dion 1973; Kramer and Brewer 1984; De Cremer and Van Vugt 1999),[10] contact and mutual promising (Dawes, Orbell and van de Kragt 1990),[11] common fate (Rabbie and Horwitz 1969; Brewer and Kramer 1986),[12] and interdependence (Sherif et al. 1961; Blake and Mouton 1986; Cookson 2000). In most of this work the basic task is choosing in a social dilemma, but sometimes it is *n*-person Chicken;[13,14] usually the group is small (from three to about ten), and the task is repeated for the same group, perhaps eight or ten times.[15] In a few cases (for example, Kramer and Brewer 1984; Bornstein, Gneezy, and Nagel 2002) it was arranged that the experimental group perceived the presence of an out-group; this also raised the rate of cooperation.[16]

4. Endogenous Group Identification in Games

4.1 *Interdependence and Common Interests as Primers of Group Identification*

I am interested in the effects of spontaneous framing in games. Group identification is a framing phenomenon. So I am interested in the role of spontaneous group identification in decision-making. (One of my main arguments will be that in games with certain features we tend spontaneously to frame the situation in a way that involves reidentifying ourselves as members of the player-group, and that this has profound consequences for behaviour and in particular for cooperative behaviour.) Thus what suits my purpose is to tease out from the above-recounted intricate story that part of it which concerns features of decision situations that tend to produce group identification. The stuck lift-passengers and the informal workgroup are the groups for me.

As we have seen, psychologists have suggested a number of conditions which tend to promote a sense of group identity, including falling within the same natural social boundary (such as all being students, or elderly, or residents of the same town), or the same artificial category (such as being overestimators of the number of dots on a screen), meeting, having common interests, being subject to a common fate, shared experience, interdependence, and a competing outside group. Two of these, 'interdependence' and 'common interests', are of particular interest for our enquiry, because they connect group identity to characteristics of decision problems.

People use both 'interdependence' and 'common interests' to mean several different things,[17] and sometimes slide from one to the other, or deliberately use them in a fuzzy way,[18] so let us try to make some distinctions and to formulate a clear version of the hypothesis in social psychology that these factors induce group identification. The formulation must if possible fit the examples most often given in the psychological literature to illustrate situations in which there is interdependence or common interests, which are preeminently Dilemmas of different sorts.[19]

First, common interests. To say that two agents, P1 and P2, have common interests is deeply ambiguous, because this statement does not tell us in what matters they have common interests, and in particular whether, given their options for action, they have identical interests over all the possible outcomes, or only some; it does not even make clear whether they have any interest-relevant actions at all. The characterization is also ambiguous about the agents' perceptions and beliefs: does 'having common interests' imply that the agents see that they have these interests in common? I propose to eliminate these ambiguities, at the cost of a small departure from the vernacular. To say that P1 and P2 have common interests presupposes that there are at least two possible states of affairs, s and s^*, in one of which, s^*, the interests of both are better served than in the other.[20] My definition will be of common interests regarding a pair of states. I will generalize this to certain sets of states, when both interests would be better served in one set than in the other. I will also build into the definition the condition that the agents have mutual knowledge of the facts in question.

My approach means that saying that P1 and P2 have a common interest is completely consistent with their also having some very non-common interests. There may be common interest in one matter and opposed interest in another. This problem appears in all 'mixed-motive' games. For example, in Prisoner's Dilemma and bargaining games, there is both a common interest in reaching the Pareto frontier and a sharp conflict of interest between reaching different points on it. But the problem is not confined to mixed-motive games. It is clear that interests are perfectly harmonious in pure coordination games, but also that the strength of the common interest in coordinating is greater when the coordination payoffs are high or the losses from coordination failure are great. Only at the other end of the scale of harmony, zero-sum games, where the degree of common interest is, intuitively, zero, does the matter seem clear.[21]

Let us say that P1 and P2 have *a common interest in* s^* over s if they mutually know that both prefer s^* to s.[22] This definition can easily be extended to the case of more than two states ordered in the same way by the two agents. It can also be extended in another way, less obvious but, as it turns out, important for understanding some benchmark interactions. If S^* is a set of states all of which are strictly Pareto-superior for P1 and P2 to a state s, I will say that P1 and P2 have a common interest in S^* over s. I'll say this even though (necessarily) interests are opposed in comparisons between states within S^*. An example of this sort of case is a bargaining situation: in my terminology, the agents have a common interest in reaching some bargain or other.[23] Similarly, I'll say P1 and P2 have a common interest in s^* over S if s^* is Pareto-superior to every state in S. 'Common-interest games', including Hi-Los, are examples of this pattern in which S is the set of all feasible outcomes other than s^*.

There are four types of cases of common interests in a state s^* over a state s which differ in whether and, if so, how the agents can influence the state. In case 1 neither has any influence, in case 2 each can bring about either state independently of what the other does, and in case 3 only one can do so. In case 4, P1 and P2 can between them bring about s^*, but only by an appropriate combination. Each has a plural option set; call these A_1,A_2. Only options from a strict subset of A_1 combined with options from a strict subset of A_2 produce s^* rather than s (or, more generally, something in S^* rather than something in S). For example, P1 has just two options, a^* and a, and P2 just two options, $b^*$and b; (a^*,b^*) yields s^* and the other profiles yield Pareto-worse outcomes. In case 1, P1 and P2 have a *common fate*: things will turn out either well for both or badly for both, but it is out of their hands. Thus case 1 does not describe a feature of a decision situation. Cases 2 and 3 play no part in the group identity literature.

I come now to the hypothesis that perceived 'interdependence' prompts group identification, the *interdependence hypothesis.* A typical informal definition is having common interests, or common goals, that can only be achieved together. At first blush this seems to mean that among all the feasible outcomes there is one—call it s^*—that both P1 and P2 rank highest, and that it can only be brought about by each acting in a particular way. This is the pattern in a 'common-interest game'. But it cannot be what is intended by advocates of the interdependence hypothesis, because the Prisoner's Dilemma doesn't fit it. In the Dilemma, (C,C) is *not* preferred by both agents to all other feasible outcomes: P1 prefers (D,C) to it, and P2 prefers (C,D) to it. It seems that the 'common interest that can only be achieved together' cannot be cashed out as a preference for a certain outcome over all others.

Let's say that in case 4 the agents P1 and P2 have *copower* for s^* over s (or, for S^* over S). Having copower for s^* over s when there is a common interest in s^* over s is one element of what I think is intended in the literature by saying P1 and P2 are interdependent. So my notions of interdependence and common interest are linked: my notion of interdependence includes copower, which presupposes that there is a common interest.

But there is another element. There are many hints in the literature that what produces group identification is that P1 and P2 are interdependent in the strong sense that they depend upon each other, that is, they perceive that they will do well only if the other does something that does not seem to be assured. I will say that the agents are *strongly interdependent* if this is so. The overwhelmingly most frequent example of a scenario in which a sense of interdependence is said to promote group identification is certainly a case of strong interdependence. It is the Prisoner's Dilemma. Each will do well only if the other does C, and C seems to each very far from assured. Evolutionary considerations also support the gloss that it is strong interdependence, if any kind, that is a spur to group identification. We shall see in the next chapter that the mechanism of group identification has particular adaptive value precisely when individual reasoning fails to realize common interests. Henceforth I will interpret the interdependence hypothesis to concern strong interdependence.

Implicitly, if a perception of interdependence can lead agents to group identify, they currently self-identify as individuals. Hence they deliberate in the ways that are characteristic of individualistic self-identity. We are led to conclude that the interdependence hypothesis concerns situations in which s^* is not assured by individualistic decision-making, and P1 and P2 perceive that it is not. For the sake of definiteness, let us interpret 'individualistic decision-making' as decision-making as given in (standard) game theory.[24] We can now give the interdependence hypothesis explicit game-theoretic form. Write sol(G) for the solution set of game *G* in standard game theory. A game *G* has feature I if:

(I) For some S, S^* the players have common interest in, and copower for, S^* over S, and sol(G) contains outcomes in S.

Then the interdependence hypothesis can be expressed thus:

(IH) Group identification is stimulated by the perception of feature I.

To see what feature I amounts to, consider a case in which individualistic reasoning gives no trouble—in which P1 and P2 have a common interest in, and copower for, s^* over s, but s^* will come about anyway through individualistic reasoning. Such is the situation when, for example, each of us wants as many of us as possible to do one act rather than another, and doing it is has low personal cost. For example, we both want a resolution to be carried at a meeting and think that the more of us who turn out the better are the chances. This is illustrated in figure 2.2, in which T is turning out, U is not turning out, turning out has a cost of 1 and each player gets a benefit of 10 for every player who turns out. Individual rationality ensures that we will all turn out, since doing so dominates.

If a game has feature I, individualistic reasoning cannot be relied upon to deliver s^*. Putting it the other way a round, individual decision-making makes possible an outcome that is Pareto-dominated by another feasible outcome. Feature I is possessed by most of the landmark games of the theory of cooperation: not only dilemmas, but also Stag Hunts, Hi-Los, Battles of the Sexes,[25] and any bargaining game

		Player 2	
		T	*U*
Player 1	*T*	19, 19	9, 10
	U	10, 9	0, 0

Figure 2.2. Turning out

in which breakdown is a possibility.[26] The fact that some games have feature I is the same as the fact that individual rationality can be collectively inefficient or 'group irrational'. For if a game has I then there is an outcome *s* which is admitted as individually rational, while there is some other outcome in which all could do better.

4.2 *How Likely Is Group Identification?*

Saying that group identification is stimulated means that there is a positive stimulus to group identification, and therefore that the probability of group identification is increased. It implies nothing about the size of the positive stimulus or of the increase in the probability. At this stage almost nothing is known, or even ventured, about this question. Write *g* for the probability of group identification. The picture is that perceiving I raises *g*, but other perceptions may do so too, and still other perceptions may militate against group identification and so tend to lower *g*. In a Prisoner's Dilemma, players might see only, or most powerfully, the feature of common interest and reciprocal dependence which lie in the payoffs on the main diagonal. But they might see the problem in other ways. For example, someone might be struck by the thought that her coplayer is in a position to double-cross her by playing D in the expectation that she will play C. This perceived feature might inhibit group identification.

The interdependence hypothesis expresses a conditional: *if* a player perceives feature I then this perception stimulates group identification in her. What is the point of the condition that the agent *perceives* feature I, and how likely is it to be met? In the main tradition of game theory this condition seems supererogatory, for it is guaranteed by the assumptions of that tradition: if a game has this feature every player knows it, for the feature is a logical consequence of what she is assumed in the definition of the game to know, and players are assumed to be 'logically omniscient'. But the assumption of logical omniscience is very strong, even in the context of a game between two players with only a few options each. Some logical consequences of what game players know are far from obvious. In general there are two ways in which a consequence *Q* of a set *P* of premises may fail to be obvious. One, recognized in the literature on logical myopia, is that it may be difficult to grasp how *Q* follows from *P*, for example because the derivation is complex;[27] this is an effect of bounds on human reasoning ability.

The other is a framing effect: that *Q* follows is not recognized because *Q* is not in the agent's frame. It is easy enough to see that *Q* follows once the question is raised in the agent's mind; but it may not be. If not, the failure to perceive Q is an example of conceptual myopia, an effect of the boundedness of human frames. We have already seen [in sections 3 and 4 of the introduction, material which would have been covered in MB's chapter III] that players' conceptualizations of the games with which they are presented—the sets of features that they perceive—are variable and incomplete. Conceptual myopia means that the presence of feature I may be less noticeable in some games that have it than in others. If this is so, it follows that the priming of group identity through perceiving I is more likely in some games than others.

The outcome for g turns on the relative salience of the different features, and on their tendencies, if and when perceived, to stimulate or inhibit group identity. In a Hi-Lo with $a > 0$, $b = 0$, [that is, a payoff to (A,A) of (a,a) and a payoff to (B,B) of (0,0)], the common interest forces itself very strongly on the attention. In general in common-interest games with feature I, it may be that if the Pareto-optimal payoff is very large, or some other possible payoff is very large and negative, this increases the salience of the common interest and of the interdependence. It may also be that there is a positive relationship between salience and effectiveness: when a feature tending to promote self-identity is highly salient, then if and when it is noticed it is also highly effective. These are empirical speculations; their investigation will be an important part of the future development of the theory of group action put forward in this book.

4.3 The Goals of Endogenous Groups

A fundamental element in the notion of a psychological group is the goal, objective or perceived interest of that group. A game-theoretic treatment of agents who may group-identify must therefore specify the membership of a group, but also determine a payoff function to represent the group objective. It will be convenient at this stage to denote a group by a pair $G = (S,U)$, where S is a set of agents and U a group utility or payoff function. In the case of endogenous group identification by the players in a game, the group membership is clearly the player set P, or perhaps some superset of P. But what this identification process implies about what these players want as a group, about what U is, is less clear.

It is implicit in the literature on the identifying effect of common interest that, in some sense, the individuals see the group objective as the furtherance of that common interest; for example, in a lifeboat the group objective is just everyone's objective—that the boat reach land as quickly as possible. In cases of perfect harmony of common interest, the shared ranking of outcomes is also the group ranking. If the game is a Hi-Lo, it is clear that the group payoff is just the shared individual payoff. But what if, though there is some commonality of interest, it is not complete? If perchance the players of a Dilemma group identify, what will they want to achieve as a group? How, for example, will the group feel about (C,D) and (D,C)?

Meanwhile, what we can cull from the literature is this. Given a game Γ with a payoff structure Π, if group identification takes place it will be identification in a group $G = (P,U)$, where U expresses the common interest embedded in Π, and 'expressing the common interest embedded in Π' satisfies two conditions. The first of these, which we may call the Unanimity condition, is that if u_1 and u_2 agree, then U agrees with them (to 'agree' is to have same value on every profile in the profile set). The second I shall call Paretianness. It is that if p is weakly Pareto-superior to p' then $U(p) \geqslant U(p')$. These are not really two conditions but one, since Paretianness implies Unanimity. Of course there is a vast range of payoff functions U that have Paretianness. For example, both the utilitarian function and the weighted utilitarian functions [that is, functions that are additive combinations of the individual players' payoffs, but in which different players may be given unequal weight] that put all the weight on a single player, are Paretian. The question is: how are players identifying with one another as a group for the perceived common interest likely to see the group's interest in furthering different members' individual aims when these conflict? It is intuitively clear that they will not see it as directed towards a single player's individual benefits. A testable hypothesis is that, in circumstances in which nothing is perceived by individual members about other individual members beyond the facts recorded in a bare game representation, principles of fairness such as those of Nash's axiomatic bargaining theory will be embedded in U.[28]

5. Who Am We?

The notion and theory of group identity have been around in social science at least since Durkheim (1893), that of *personal identity* in philosophy and psychology for much longer. I suggest a partial unification of the two notions, which goes beyond the linkage we find in self-categorization theory. That theory points out that, loosely, personhood is to some extent constituted by group memberships: membership of a gender, race, class, family, profession, supporters' club, fan of this or that singer, driver of this or that car, watcher of this or that cult television series. Personhood is the resultant, to the extent that it is so constituted, of a set of group identities; more exactly, the person is defined by the intersection of her group identities. But it is only to some extent, since there are plenty of person-defining features which do not correspond to group memberships.

My suggestion assimilates personhood to group identity more directly and more fully. Since a person is temporally extended, we can represent her as a sequence of adjacent, linked segments—'time-slices', 'dated subpersons', or what have you. Each of these parts has the essential characteristics of an agent and, in particular, is the locus of the person's thinking about what to do, now and later. There is a considerable literature (for example, Strotz 1955–56; Elster 1979; Ainslie 1992;) which uses this multi-agent model of a person in order to study questions in the theory of dynamic choice—notably how and why people make plans, and what explains why they stick to them in the face of temptation. In this literature the agent-like parts are treated as the players in an n-player game. This model has its merits, as we shall see later. But it also has a glaring deficiency—it fails to explain the sense of personhood which a person has (in each time period) when she thinks inwardly.

We can partially make good this deficiency and, I will argue later,[29] illuminate some puzzles in the theory of dynamic choice, by thinking of subjective personal identity as a special case of group identity in which each temporal agent in the sequence entifies the whole sequence, and conceives itself as a member of this entity. It has thoughts like 'I am Zelda'. This 'am' raises a puzzle. How can it be true that something (a subperson) that *is* part of a person *is* a person? The answer is, I think, that the theoretical entity picked out by the first 'is' (the agent) is engaging in the talk characteristic of group identifiers. Thinking that she is Zelda is analogous to thinking that we are Zanzibar.

6. Stocktaking

We have seen weighty evidence of a psychological phenomenon, group identification, which has a number of characteristic causes and effects, and is so-called because of its feature that the subject of the phenomenon thinks of herself first and foremost as a member of the group. It appears from Tajfel's work to be unnecessary that the group should have goals or powers to affect things, but, on the other hand, if there are common interests or interdependence then group identification is more likely. I suggested a precise spelling out of these notions, 'strong interdependence', which is present in an interaction if agents thinking as individuals might fail to realize a common interest. The laboratory test literature we surveyed in section 3.3 shows that, if group identification does take place, it enables members of the group to achieve their common interest, in other words, to cooperate.

How and why group identification, if it does occur, might enable cooperation is a part of the picture that is still blank. If the potential of group-identification to explain cooperation is to be fulfilled, we need to know, in detail, what the deliberations are like that people engage in when they group-identify. We have a good idea of part of the answer: there is, most agree, a payoff transformation. But having the group interest at heart does not, as we have seen, suffice to explain the facts about Hi-Lo or cooperation in other common-interest games. I will suggest in the next part of the book that something more happens, seemingly bizarre, but in fact most natural given the nature of group identification. This something is 'agency transformation'. The key to my explanation is that agency transformation involves not only a transformation of payoff but also a transformation of reasoning.

Among the claims about human groups I have been surveying, of special importance for my own argument in this book are three: (i) we frame ourselves as members of groups; (ii) the goal of the group in this framing of self need not agree with the person's goals under her individual framing of herself, but perceived agreement of individual goals among a set of individuals favours framing as members of a group with this common goal; and (iii) the group framing tends to issue in efficient cooperation for the group goal. Together these three claims constitute a core subset of features of group identification (framing, common purpose, cooperation), which we may dub the *core mechanism*. They are at best bizarre claims from a standard perspective, and downright contrary to the principles of hard-line rational choice theory. In the next chapter I will try to display them in a new light, by asking whether we would expect them to be part of human nature.

Notes

1. There is evidence that these properties are more salient after group identification of the self. Smith and Henry (1996) found that identification with comembers of a group enhances the accessibility of shared characteristics: reaction times in self-rating tasks were faster for traits previously deemed shared.

2*. MB planned to cite experimental evidence of each of these.

3. In the early twentieth century many social theorists accepted a very strong conception of the group as agent, deriving from Rousseau and Hegel, which attributed individual-like mental activities and states to the group, even sentience (Sober and Wilson 1998).

4*. MB planned to write a paragraph here on why members of nationalistic movements are ready to sacrifice their lives for the cause.

5. I know of no experimental evidence of this, so I'm sticking my neck out here. But I think I'm quite safe. First, introspective evidence persuades me that when I'm identifying with a group, when I speak I refer to it as 'us'. Second, Gilbert (1989) has stressed that there is a sense of 'we' that is in common use, and is used only when the referent is a social group. A social group is, for Gilbert, that which has joint intentions and does things together. I shall argue later that joint intentions are best explained as arising from group identity.

6. Depersonalization is a cluster of psychological phenomena including the breakdown of normal self-monitoring and self-restraint, heightened emotions, and inability to plan rationally (Diener 1977).

7. Emotional contagion involves the tendency to fall into the same mood as other group members. This can work by what seems a causally bizarre route, 'self-feedback': a group member tends first to mimic the outward signs of others' moods, their facial, postural, and vocal behaviour; once she has adopted these, it has been shown (by measuring muscular, visceral and glandular responses) that she begins to feel the emotion that produced them in others (Hatfield, Cacioppo and Rapson 1994).

8. We observe that people who want to belong invent and constantly use 'in' words, jokes, myths and nonverbal pieces of behaviour. Their use of them may be not only signal-giving evidence that they belong, but also Gricean speech acts affirming their wish to do so (Grice 1989).

9. Cookson gave subjects classic public good contribution decisions. Each of a set of four subjects was given 400 points (worth money) and invited to pay in ('contribute') any number of these to a pot. The pot gets doubled and shared out, so that each subject gets back half the total contribution. Formally, subject i's money payoff ($i = 1, \ldots, 4$) is

$$y_i = 400 - c_i + \sum_{j=1}^{4} c_j / 2$$

This is a generalized Prisoner's Dilemma with four players and 401 levels of 'cooperativeness'. If all contribute 400, all finish with 800. For each, the dominant choice of contribution is zero, since whatever others' contributions may be, each unit you contribute reduces your net payoff by 0.5. But if all contribute zero, all finish with 400. Subjects played blocks of eight rounds of this game, the second and third blocks being preceded by one of two comprehension tasks, task *I* and task *W*. In task *I* a subject had to write down how much she would get as she varied her individual contribution; in task *W*, how much each person in the group would get if they all contributed the same number, varying this number. These tasks were designed to prime individualistic and group frames by their focus on individual and common acts and gains. But the content of the *W* tasks may have also made group identity more salient by highlighting interdependence. W comprehension tasks were followed by more cooperation than were *I* tasks.

10. For example, Kramer and Brewer (1984) gave groups of three subjects a repeated replenishable but exhaustible resource task, and observed higher rates of restraint when a subject group, knowing that they all lived in Santa Barbara, were told that the experimenters were interested in the behaviour of Santa Barbara residents.

11. Dawes, van der Kragt and Orbell (1988) studied the effect on cooperation of allowing the subjects to talk to each other freely for ten minutes before the decision task, including discussing the task and saying how they would choose. The task was a step-level public good problem: the five players all received a bonus if at least three made a contribution. A marked positive effect of preplay talk was observed. Since promising was allowed, it was unclear whether the effect was due to the promising (for example, making promises strategically to induce reciprocal promises, together with moral scruples in those who had made promises); or to group identification, produced by, for example, mutual acquaintance and a shared experience. In 1990 Dawes, Orbell and van de Kragt partially resolved this issue: they made the surprising discovery that contribution rates were markedly higher when everyone had promised than when promising had fallen short of being universal even by one. Their interpretation was that universality may cause, and/or be caused by, group identification.

12. Brewer and Kramer (1986) presented subjects with a public good problem and a commons dilemma, each with either an individualistic or group identity manipulation. In the commons dilemma, subjects took less for themselves (44 rather than 74) when collective identity was salient, but this effect was much stronger when the resource had become severely depleted. In the public good task, they behaved slightly more cooperatively in the group identity condition, when the group was small (8), though not when it was large (32). To induce group identity the experimenters told subjects the value of each point earned would be 2 cents or 1 cent according to whether the computer's random number generator yielded an odd or even number; for individual identity they were told that the value would be determined in this way independently for each subject in a session. In addition, the subjects were referred to as 'group members' and 'individuals' in the instructions for the two conditions.

13*. In the classic two-person Chicken game, each player chooses between a submissive strategy (*dove*) and an aggressive strategy (*hawk*). Typical payoffs are shown in figure 2.a. The prototype of Chicken is the game of bravado played by teenage boys in which, for example, the winner is the last boy to get out of the way of an oncoming train.

		Player 2	
		dove	*hawk*
Player 1	*dove*	1, 1	0, 2
	hawk	2, 0	−2, −2

Figure 2.a. Two-person Chicken game

14. In van de Kragt et al. (1986), one task was a step-level public good contribution problem; such problems are *n*-person Chickens. In the basic step-level problem with *n* players, each can contribute her whole endowment or none, and if at least *m* of the *n* contribute, all *n* receive the public good. In the van de Kragt experiment, if five out of nine contributed $5, all nine received a $10 bonus. There are many equilibria of *n*-person Chicken with these parameters, including contributing nothing. There are 126 Pareto-optimal equilibria in which exactly five contribute, and a severe coordination problem for players who aim for one of these. To block discussion of which five would contribute, *m* was not revealed until after the discussion, so what was played was an incomplete information version of *n*-person Chicken.

15*. MB planned to insert a note on the number of, and effects of, repetitions.

16. Bornstein, Gneezy and Nagel (2002) got a group of seven subjects to play the Van Huyck, Battalio and Beil (1990) minimum-effort game. [Each individual chooses an 'effort' level from 1 to 7. Each individual's payoff is decreasing in his effort level and increasing in the minimum effort level of the group. The best outcome for each individual is the one in which everyone chooses 7, but the best response is to match the lowest effort level chosen by any other group member. This game can be thought of as Stag Hunt with many players and many strategies.] In one condition they play normally. In the other there is another group playing the same game; the group with the higher minimum effort gets standard payoffs, the other nothing. After each of the ten rounds, each group is told the minimum effort of both groups in the last round. This treatment improved collective efficiency. The authors conjecture that the presence of the out-group makes winning against the out-group salient and produces focality. My own interpretation is that group identity enters twice: first (as is implicit in this conjecture), the players frame the situation as an intergroup one in which they are members of one of the groups; second, their intergroup conception generates a perceived group goal, that of doing better than the out-group.

17. For example, the standard meaning of 'interdependence' in game theory is simply that the effects of each agent's actions depend on those of others; this is true in all games, including zero-sum games, where no-one would suggest that it arouses a sense of group identity.

18. This is not a criticism. In describing psychological states and processes, rather than modelling them as game theorists do, it is proper to respect the complex, shifting, multivocal nature of these states and processes.

19*. MB planned to cite experiments on group identity in social dilemmas.

20. The 'states' s and s^* here are not elementary states or possible words, but events, such as 'Brazil wins the semifinal'.

21. Zizzo and Tan (2002) propose a general measure of the 'harmony' of a game, based on the correlation between the players' payoffs across outcomes, and so having maximal value in any pure coordination game and minimal value in any zero-sum game. Then index captures a number of plausible axioms for 'is a more harmonious game than', such as independence of linear transformations of the utility functions, and a 'common fate' axiom saying that (unless the game is already a pure coordination game) expanding the game in

such a way that it has an extra possible outcome that is very good for all (or very bad for all) increases harmony.

22*. MB planned a note defending the substitution of preference for interest and to foreshadow his discussion (in chapter 4 on team reasoning) of substantive goals, such as winning the cup or making your children happy. On this see also the editors' discussion of payoffs in section 2 of the introduction.

23. The bargaining situation is one in which bargains are all Pareto-superior to breakdown.

24. Of course this is not the only possible way that people who self-identify as individuals might decide—or try to decide, or put off deciding. The game theory scenario is one in which agents actively seek to determine what to do in accordance with certain principles, and complete their deliberation in good time. In a scenario far from game theory, an agent confronting an interactive situation might be taking no active deliberative steps and just letting things take their course, when struck by the thought that by doing a^*,b^* respectively the agents could bring about a good outcome (s^*) for both. In the Unlocking example of chapter 1, section 1 (Example 9), we would just sit there, Micawberishly hoping that somehow the mess would get sorted out. In a second, an agent might be engaged in deliberation but not yet have found a convincing argument for any of her options when struck by the a^*-b^*-s^* thought. My interpretation of the interdependence hypothesis, in which I take individualistic reasoning to be game-theoretic, could easily be adapted to embed any other model of individual decision-making; there would be strong interdependence if, on that model, s^* were not assured. Prisoner's Dilemmas would come out as having feature *I* as long as the model allowed (D,D) as a possible outcome.

25*. The Battle of the Sexes game is described in note 28 to chapter 3.

26. It isn't true of Chicken, as the only outcome that isn't Pareto-optimal is (Hawk, Hawk), and in standard versions of game theory this isn't a solution.

27*. MB planned to cite literature on logical myopia.

28*. Nash (1950) models a two-person *bargaining problem* as a closed, bounded and convex set X of feasible payoff pairs (u_1,u_2) for the two players, one of these pairs, (u_1',u_2'), being specified as the one that will come about if the players fail to agree. He defines a *bargaining solution* as a rule which, for every bargaining problem, picks out a unique element of X. He presents a particular bargaining solution, namely the rule that selects the element (u_1^*,u_2^*) of X that maximizes the value of $(u_1^*-u_1')(u_2^*-u_2')$, and shows that this solution is characterized by a set of simple axioms. For an exposition of Nash's bargaining theory, see Binmore (1992, pp. 180–191).

29*. MB is referring to the chapter on 'the person as a team' that he originally intended to include in the book, but later decided to drop. The editors discuss MB's ideas about personhood in section 7 of the conclusion.

Chapter Three

The Evolution of Group Action

1. Evolutionary Explanation

Explaining the evolution of any human behaviour trait (say, a tendency to play C in Prisoner's Dilemmas) raises three questions. The first is the *behaviour selection question*: why did this trait, rather than some other, get selected by natural selection? Answering this involves giving details of the selection process, and saying what made the disposition confer fitness in the ecology in which selection took place. But now note that 'When a behavior evolves, a proximate mechanism also must evolve that allows the organism to produce the target behavior. Ivy plants grow toward the light. This is a behavior, broadly construed. For phototropism to evolve, there must be some mechanism inside of ivy plants that causes them to grow in one direction rather than in another' (Sober and Wilson 1998, pp. 199–200). This raises the second question, the *production question*: how is the behaviour produced within the individual—what is the 'proximate mechanism'? In the human case, the interest is often in a psychological mechanism: we ask what perceptual, affective and cognitive processes issue in the behaviour. Finally, note that these processes must also have evolved, so an answer to the second question brings a third: why did this proximate mechanism evolve rather than some other that could have produced the same behaviour? This is the *mechanism selection question*.[1]

All this concerns a certain way of *explaining* a trait T. And explanation is all one needs, for theoretical completeness, if T is an unproblematic fact. But it may happen that T is disputed. In this case the import of the evolutionary account of the causes of T is different. Instead of explaining T, it helps to render T plausible; it adds to the evidence on the side of the claim that people have the trait T. It is evidence against T if no such account can be found—if it looks as though T could not have evolved; conversely, it is evidence for T if T (and a proximate mechanism for it) are shown to be adaptive. However, there isn't a sharp line between explaining T and rendering T plausible, since if the account does the latter well enough to lead us to accept T, we then require explanation, and the same account serves this function too.

It might be thought that if T specifies behaviour—rather than, say, motivations or other hidden states or processes—then there can be no problem about whether man possesses T: you just look. But this is false. In the first place, T might specify behaviour in the Pleistocene epoch, such as using stone tools to cut up meat, of which we can only have indirect evidence. But even if T is a claimed behavioural trait in modern man, we cannot just look and see. Behavioural traits are dispositions to act in a certain way in a certain context, and even if the act is unproblematically observable, identifying the context is usually not, since we cannot directly observe the whole of the context. Suppose, for example, that we observe a piece of behaviour in individuals that results in their getting less food and others' getting more—say, proffering the food. We cannot observe, then and there, whether they will or will not repeatedly find themselves involved in food division with the same other individuals. In just this way the claim that humans frequently make choices which are cases of C-choices in Prisoner's Dilemmas has been disputed, and the observed acts have been reinterpreted as C-choices in 'stage-games' of a repeated Dilemma (Trivers 1971).

2. What Needs Explaining

We have weighty evidence that human beings are cooperators. Some of this evidence concerns ancient humans, of the Pleistocene period, during which we were hunter-gatherers, before the rise of agriculture round 10,000 B.C.E. and is inferred from the fossil record and other data. Some of the evidence is contemporary, and ranges from surviving hunter-gatherer societies to modern families, sports teams, work-groups in industrial corporations and Internet world action groups. The early evidence also indicates cooperation in varied settings and with varied structures of interdependence and payoff. The latter-day evidence of cooperation is mostly for small unorganized groups, the early evidence mostly for larger groups, often with some degree of internal organization. Nevertheless, there is considerable overlap between the two bodies of evidence. Importantly, both relate to interactions of the same game-theoretic *genera*.

In neither case are the claimed tendencies to cooperate universal. For early times much has been written about failures to cooperate and systems of penalties against their perpetrators. For our own time we have careful measurements of rates of cooperation in specific games, such as Dilemmas, and know that uncoerced cooperation rates are only moderate; they are thought to vary with personality types, as distinguished by 'social value orientation' (Offerman, Sonnemans and Schram, 1996): there appears to be a polymorphism of more, and less, cooperative traits. Cooperation is not only *by* only some; it is also *with* only some. The evidence, early and late, is that it can be discriminatory: that is, there are cooperative opportunities in a large group, but the cooperation that takes place is among members of a proper subgroup. Moreover, cooperation can damage the interests of people outside the cooperating subgroup, and this damage can be inflicted knowingly. With regard to cooperativeness, the human race appears to be a curate's egg.

The evidence of cooperation for the early period is of special interest for human evolutionary theory. There is a presumption that behaviour a quarter of a million years ago was largely driven directly by natural selection, so for the early *Homo sapiens* period it is easier to get an explanation from pure natural selection rather than having to deal with a mixture of natural and cultural selection. And if we do conclude that a trait T was indeed naturally selected at that stage in our evolution, we can happily infer that T also characterizes modern man, even though today's environment may be much different from the one for which it was selected.[2]

There is wide agreement that the life of early humans was characterized by fruitful group activity (Caporael 1995). It developed long before the emergence of agriculture around 10,000 B.C.E., being at least as old as the earliest man, *Homo habilis*, who appeared about two million years ago. The activities included defence against predators such as leopards and hyenas, deer hunting, handling large carcasses, foraging, food sharing, and grooming (Mithen 1996, pp. 107, 128, 133). We can be fairly sure of this and of some of the features of the groups, such as their size, both from archaeological evidence (such as a *Homo habilis* skull bearing leopard tooth-marks (Brain 1981)) and from observation of latter-day hunter-gatherer peoples (Wilson 1998). Living groups varied considerably in size: early humans (those living before *Homo sapiens sapiens* appeared round 100,000 B.C.E.) lived in small groups in wooded environments and in large groups on open tundras; in a given place they apparently needed to adjust group size constantly as climatic conditions changed. At a given time and place different activities were carried out in groups of different sizes which formed a nested hierarchy, so that a given individual would have multiple group memberships. Caporael (1995) distinguishes four levels: the dyad (exemplified by parent and child, or a couple), the family or workgroup or team of five or so, the face-to-face group or 'deme' of around thirty, and the macroband of a few hundreds. Teams hunted and foraged; demes moved from place to place and coordinated the teams; macrobands gathered seasonally and exchanged resources.

The groups in which our ancestors functioned differ from each other in size, but also in relatedness[3] and, importantly for us, in 'organizational form'. There is no doubt that as man learned to live in larger and larger groups, organizational forms proliferated and more complex forms emerged. But to illustrate the notion of organizational form, the small-group behaviour of primates suffices. Boesch and Boesch (1989) classify group behaviour in Tai and Gombe chimpanzees into four forms: (i) *similarity*: all perform similar actions, not functionally related in space or time; (ii) *synchrony*: similarity plus an attempt to synchronize; (iii) *coordination*: similarity plus an attempt to relate them both in time and in space; and (iv) *collaboration*: different individuals performing complementary actions (for example, some driving the prey, some blocking escape). The attempts to relate actions are made by, for example, signalling intentions to change direction. Whether chimpanzees use collaboration is strongly correlated with how many there are in the group on a given occasion.[4]

After a successful hunt there is a food-sharing episode. The first stage of this is also partially organized, in another form, which I shall call 'simple direction' in chapter 4. If the prey is large, it is divided between two or three individuals by a single individual (whose identity depends on preestablished rank). What happens thereafter appears to be an unorganized n-player game in which individuals have to choose, roughly speaking, how well to behave.[5]

The gross fact to be explained is that early man managed to function well in groups, by doing things that we are inclined to call 'cooperating' in situations in which individuals are interdependent. But what is the fact to be explained *more exactly*? What is it for a group to 'function well', or for its members to 'cooperate'? To do well, of course, is to do well in terms of fitness. But beyond this the formulation is vague and ambiguous.[6] We need to make it more precise if we are to bring this hypothesis about ancient behaviour to bear on the explanation of specific forms of modern group behaviour. In doing this we may have to sacrifice some generality but, I think, not too much. The reformulation should work for the most frequently cited examples in the literature, capture the essential features of the situations—interdependence and doing well—and be faithful to the spirit of the literature. The examples we find—food sharing, success in group hunting, mutual defence—include interactions that are, game theoretically, of two broad kinds: they are either of the Dilemma family or common-interest encounters. In the case of the former, groups are said to have functioned well or cooperated if they manifested much C behaviour. In the case of the latter, if they tended to realize the common interest. A formulation of the notion of 'doing well as a group' that covers both these cases is achieving high total fitness. Moreover, it is apt to say that the group 'cooperated' if it did this.[7] So I shall work with the following (sufficiently) precise formulation of the thing to be explained. Let us call a group *successful* if its members' behaviour maximizes the sum of fitnesses. The explanandum is then just that early man participated in successful groups.

A simple example of a group in this theory is a pair of hunters who face a two-player Stag Hunt in fitnesses. Figure 3.1 shows the payoff matrix in fitness gains (unit fitness represents an expected number of offspring of 0.01). The matrix has this interpretation. Each player is of one of two types; she either has the S trait (has a hardwired disposition to hunt the stag) or the R trait (is hardwired to hunt for rabbit). If both players have the S trait then both finish up with two extra units of fitness; if player 1 has the S trait and player 2 the R trait, 1's fitness declines by one and 2's increases by one; and so on. Another example of a successful group is a deme of twenty individuals who have a carcass to last them for a certain period and so face a twenty-player repeated Social Dilemma, and who all exercise restraint. A third example is four hunters who hear a cry meaning an easy prey has been seen and so face a four-player Hi-Lo, and all respond to the call. A fourth is like the second except that in addition to doing C or D, members may or may not do P—play their parts, at some individual cost, in a system for rewarding C and punishing D. Although the optimal (in terms of total fitness) behaviour for such a group would be for no-one to P and everyone to C anyway, the often-observed behaviour in which everyone does C and some do P is almost as good if the cost of P is low (Sober and Wilson 1998). A fifth example illustrates the notion of a group whose behaviour discriminates between members and outsiders (Sethi and Somanathan 1999). Suppose that the interacting population contains a type of individual having a distinguishing characteristic—say, freckles—and that each member of this population faces a Stag Hunt with a random member of the population. Suppose a 'freckles' always plays S if paired with another freckles, and otherwise plays R. Then all-freckles groups will be successful groups, but mixed groups and no-freckles groups will not be.

		Player 2	
		stag	*rabbit*
Player 1	*stag*	2, 2	−1, 1
	rabbit	1, −1	1, 1

Figure 3.1. Stag Hunt

Problems of many other structures were no doubt faced by our ancestors. At one moment an early *Homo* found herself in a stage-game of a repeated interaction with a known individual; at another having to choose whether or not to come to the aid of a friend challenged by a third; at another in an opportunity to hunt with a group of members of her deme; at another in a seasonal tribal meeting at the permanent waterhole (Dunbar 1993). Early man was engaged in creating the conventions of natural language (perhaps a quarter of a million years ago according to Aiello and Dunbar 1993); in brinkmanship; in hand-to-hand combat. In game-theoretical terms, each individual played a mixture of games drawn sequentially according to some stochastic process from a large set of games. That was, and is, life. Let me call this stochastic process the *life game* of the individual. Different particular encounters in the life game involved interacting with different sets of people, drawn from her own and other macrobands. In some cases the interactions were zero-sum. In many, as we have seen, they offered scope for cooperation, and we may speak of groups as successful or not.

The encounters an individual had can be classified according to their game-theoretical type—repeated Schelling coordination game, Chicken, Prisoner's Dilemma, repeated Prisoner's Dilemma, repeated Stag Hunt, public good game with random rematching, and so on. I am going to refer to the set of game types contained in the individual's life game her *ludic ecology*; and I will refer to the size and variety of this set as her *ludic diversity*.

Standard models in bio-evolutionary game theory postulate a ludic ecology having minimal ludic diversity—it contains just one game type—and that each individual has a single genetically given trait which determines behaviour in games of this single type. For example, Maynard Smith (1982) studies a model in which organisms play a one-shot two-player Chicken and have either a Hawk or a Dove trait, Sober and Wilson one in which they play a one-shot n-player Prisoner's Dilemma and have either an Altruism or a Selfishness trait, Myerson (1991) one in which they play a Stag Hunt and have either a Stag or a Rabbit trait. To be sure, writers work with bio-evolutionary models for different types of games, but never with models of ludic diversity.[8]

It is clear enough that ludic diversity was substantial both in terms of the sheer number of types of games faced and in terms of the distance between those types, and that it is desirable to develop the bio-evolutionary model in this direction.[9] To explain the fact that early man 'functioned well in groups', we need to explain why he had, for example, a disposition to choose C in Prisoner's Dilemmas, a disposition to choose S in Stag Hunts, and a disposition to choose A in Hi-Los. More abstractly, what we need to explain is a *repertoire* of dispositions: trait

T_1 in I_1, . . . trait T_n in I_n, where I_1, . . . I_n are interactions with different payoff structures that give scope for cooperation. (Since humans evolved to be highly facultative, we would expect an early human to have a correspondingly wide range of abilities, geared to different situations.)

We need to conduct such an enquiry holistically. Here is why. The theory must not only show that each of the traits would have been selected, but also, for each, must exhibit a plausible proximate mechanism: M_1 for T_1, M_2 for T_2, and so on. Now the plausibilities of these trait-generating proximate mechanisms are not independent of one another. One reason is that if there is a relatively simple mechanism M that supports several traits, this may confer an evolutionary advantage on M over other less versatile mechansims. I shall return to this point a little later on.

The diversity of the life game means that in explaining ancient group successfulness we should not confine our attention to any one game type. Yet the presence in the life game of Dilemmas merits special attention. Evidence that groups facing such problems did cooperate (for example, many surviving hunter-gatherer peoples engage in efficient food sharing) creates a famous paradox in evolutionary biology. On the one hand, cooperative behaviour in fitness Dilemmas has evolved. But on the other, on the standard evolutionary model, natural selection could not possibly select for cooperative behaviour in fitness Dilemmas. For in such interactions a trait for C (restraint, say) would necessarily lose its possessor fitness relative to owners of a trait for D (grabbing).[10] Three responses to this challenge have been seen. The first is the model of Trivers (1971), which respecifies the basic interactions as repeated Dilemmas and shows that in such an ecology a trait could evolve for 'reciprocal altruism'—playing C if and only if your interactant has played C on previous occasions. The second is the 'kin selection' theory of Hamilton (1963), which respecifies the selection process by supposing that population dynamics make members of any given group tend to be genetically related. Now, a C trait, though still disadvantaging its owner, can confer outweighing fitness benefits on others who, because related, are likely to be C-traiters. In this way the fraction of C in the population can expand from generation to generation. The third is 'group selection'.

3. Between-Group Selection

Social scientists say that in a Prisoner's Dilemma the choice D is 'individually rational' and that the profile (C,C) is 'group-rational'. Intuitively, what makes (C,C) group-rational or 'rational for the group' is that the group does well. Could the fact that *the group does well in fitness* if much C behaviour goes on in it explain the early emergence of successful groups? The suggestion seems to entify the group. It also leaves a great gap in the explanatory story that runs from the forces driving individual behaviour to the group's doing well. Early proponents of 'group selection' indeed thought sloppily about such matters, helping themselves to assumptions that groups and species were functionally organized entities and that individual behaviour was driven by concern for group welfare. We how have, however, a nonsloppy version of group-selection theory (Price 1970; Hamilton 1975; Sober and Wilson 1998), committed to no strange entities or undefended assumptions. This version is methodologically individualist in the sense that the only traits that get selected are traits of individuals, and selection is of these traits. The notion of group is as thin as can be: all we need is the notion of an interacting set of individuals. The heart of the explanation the model gives of the evolution of cooperation is—to use economic terminology—that groups are sets of beneficiaries from fitness-externalities of the behaviour of individual members. The evolution of cooperation is a story of the progressive internalization of these externalities. Group-selection theory makes it possible to argue that man has by nature certain externality-generating, cooperative dispositions—dispositions which could not have evolved under individual selection processes. I now explain nonsloppy group-selection theory.

Consider a population of individuals in period t who cause genetic copies of themselves to be produced in period $t + 1$ if they score highly on some variable x (because it aids survival to breeding age and/or the likely number of offspring if this is reached). Consider a trait T which some have and some don't. The basic idea of group selection is this. Suppose that (i) the population is divided into interacting subsets, and (ii) if a subset has more T's this raises the *average* x *score in that subset*; then the fraction of T's in the population can grow between t and $t + 1$. The point here is that what makes T spread is not that having T gives *you* a high score. Indeed the main interest in the literature is about a case in which it does just the opposite. The point is that your subset's being rich in T raises the (average) score of your subset.

The classic example is the case in which the basic interaction is a Prisoner's Dilemma and T is some trait that leads to playing C. Here, even though T individuals do *worse* than non-T ones, *subsets* with more T's do *better* on average. Determining the net outcome thus requires further modelling. The underlying reason is that the action produced by T (that is, C) confers external benefits on other subset members. High-T subsets internalize a lot of this externality.

The fact that being T is worse for the individual than being not-T means that, within each subset, the fraction of T individuals declines from t to $t + 1$. This property of the selection process is described by saying that there is selection 'against T within groups' or 'at the individual level'. *But*, the fact that being T-rich is better for a subset than being T-poor means that the T-rich subsets grow faster between t and $t + 1$. To describe this property of the process we say that there is 'selection for T between groups' or 'at the group level'.

What I have said so far is short term: it concerns changes in the incidence of T in a single selection step, between t and $t + 1$. But the behaviour selection question (why did T and not another trait get selected?) is a question about the long term, about the course of evolutionary history. We have seen that the fitness of a trait T in any one generation $(t,t + 1)$ depends on on there being T-rich groups at t. If there are, the T fraction in the population in $t + 1$ may be higher than it was in t. But it will only rise from $t + 1$ to $t + 2$ if there are T-rich groups at $t + 1$. The same goes for each generation. So the key to the long-term prospects for T is the process that determines the extent to which T's get grouped with T's in each new generation.

In one benchmark case, in each new generation a given individual interacts in an n-player encounter with $n - 1$ coplayers selected randomly from the whole population. The encounter fixes its x score and thus the number of its progeny, which (in the basic case) have the same trait as it. Each of its progeny now interacts with $n - 1$ coplayers, once again selected randomly from the whole population. This dynamic grouping structure is exemplified by a process of 'constant fragmentation plus uncorrelated regrouping' (Hamilton 1975). It is the structure standardly assumed in evolutionary game theory (Friedman 1991). In another benchmark case everything is the same until the rematching stage, but rematching preserves groups in the sense that the progeny of any given group interact only with one another. This is what happens in an 'island' model, the limiting case of a population structure with high intergroup 'viscosity' (Wright 1945; Hamilton 1964; Myerson 1991); groups are 'spatially isolated multigenerational units'. These are two polar regrouping processes. In both of them groups of a sort figure, but in the first the group is ephemeral—the collection of n individuals thrown together at a stage—and in the second persistent. A third case is in between. It is called 'assortative settling' by Hamilton. After reproduction, the population is repartitioned into interacting sets, but the types of an individual's interactants are *correlated* with her own. A T-traiter is more likely to find herself with other T-traiters than with non-T-traiters.[11]

The key to the long-term outcome is the degree of correlation in the regrouping process. If regrouping is assortative, T's tend to find themselves with T's: at the start of $t + 1$ the now larger fraction of T's in the population are therefore distributed in a way that is favourable for a further increase in the fraction of T. The more like-with-like grouping there is in the population at a generation, the greater is the average fitness of T's at that generation; the less there is, the lower it is. Another way of saying this is that single-step group-level selection pressure, and hence single-step growth of T, is an increasing function of the variance in the T-richness of groups. This is shown formally in Price's equation.[12] The upshot is that T may be able to continue to grow, notwithstanding the selection pressure against it at the individual level, if, but only if, something or other brings it about that when groups form, they do so in a way that brings T-traiters together. Cooperative behaviour may be able to evolve in a dilemma-full world if, but only if, regrouping is assortative.

A natural way for this to happen is that 'the same groups' are present in each generation, and that whatever it is that makes a set of interactants at $t + 1$ be 'the same group' as a set at t is something that tends to keep its T-fraction constant. A good example is groups defined as the occupants of different spatial locations, so that sets of interactants at different times are 'the same group' as they are the occupants of the same location. If the locations are dispersed, there may be a tendency for the assortativeness condition to be met, because it is difficult for offspring to migrate into other groups. The logic here is: (1) there is a nongenetic attribute L (for example, location) which partitions the population at time t into classes $L_1, L_2, \ldots$; (2) the sets of interactants at t (the 'groups' of the theory of group selection) are just these classes (that is, if two individuals have the same L value they are in the same interaction); (3) the offspring of an individual are likely to have the same attribute value as herself.[13] But several other mechanisms for assortativeness have also been suggested, such as a tendency for high-T groups to vet potential immigrants for T-ness before admitting them (Sober and Wilson 1998).

4. Group Selection of Coordination Ability

The fact that there can be selection of a trait at group level when the basic encounter is a Dilemma is the root of an important and elegant theory about the evolution of cooperation, set out formally by Hamilton (1975) and elucidated and defended by Sober and Wilson (1998). It is important because one major kind of interaction in which humans cooperate is Dilemmas, and elegant because, prima facie, cooperation in them could not be naturally selected. But the fascination of these issues—how could cooperative behaviour emerge in a dilemma-crammed world? could it be the case that group selection did the trick?—has focussed our attention too narrowly. Recall that the problem we confront in seeking to explain the evolution of group well-functioning in man is to explain its evolution in a variety of interactions, from Dilemmas through Stag Hunts to Hi-Los.

When we turn to explaining human behaviour in interactions of non-Dilemma form, group selection still plays an essential role. It does so even in *coordination* settings. The fundamental trick of group selection of a trait T is that T-rich groups do better than T-poor groups. In the case of the Dilemma, C-rich groups do well because each C-er confers external benefits on all other members. In the case of coordination games the reason T-rich groups do better is different. Let T be a trait for the most efficient way of coordinating; then T-rich groups do well because in them efficient coordinators are teamed with other efficient coordinators. By the very nature of efficient coordination, one needs efficient *coordinators* in the plural for the efficient outcome to be realized; good coordinators are no good to anyone unless they are working with other good coordinators; it takes two to tango. The benefit from the presence of a T-traiter in this case falls on *all* the team members, including the possessor of the trait.

Example: Stag Hunt world. Consider a world in which the population is split into large groups and each individual plays a sequence of two-player Stag Hunts with a random sequence of comembers of its group. The fitness payoffs are as shown in figure 3.1 above. [A creature which always plays R has a fitness of 1, irrespective of what other members of the group play. If the proportion of S-playing creatures in the group is σ, a creature which always plays R has an expected fitness of $2\sigma - (1 - \sigma)$, that is, $3\sigma - 1$. The solid lines in figure 3.2 plot the individual fitness of R- and S-playing creatures (on the vertical axis) as functions of σ (on the horizontal). In any group in which $\sigma < 2/3$, an S creature is less fit than an R creature. So if $\sigma < 2/3$, there is selection against S at an individual level, and the fraction of S is lower in the group after breeding than before. Conversely, if $\sigma > 2/3$, there is individual selection in favour of S. These forces of selection are represented in figure 3.2 by the arrows labelled 'individual'. But now consider the average fitness of members of the group. This is $(1 - \sigma) + \sigma(3\sigma - 1)$, that is, $3\sigma^2 - 2\sigma + 1$. This expression, plotted by the dashed curve in the diagram, is at a minimum when $\sigma = 1/3$. Thus, at all values of σ greater than $1/3$, average-member fitness is bigger the bigger is σ so between groups with $\sigma > 1/3$, there is selection at the group level for groups with higher values of σ. Conversely, between groups with $\sigma < 1/3$, selection at the group level favours groups with lower values of σ. This implies the forces of group selection shown by the arrows labelled 'group'. In a population made up of groups with values of

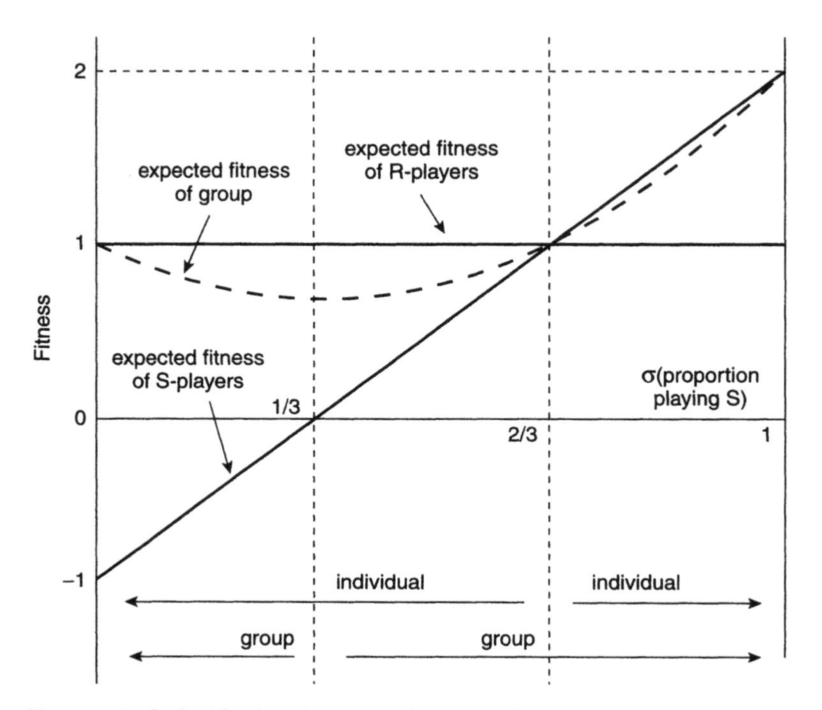

Figure 3.2. Individual and group selection in Stag Hunt

σ between 1/3 and 2/3, selection forces at the two levels are opposed.[14]] What happens depends on the population dynamics—on who gets to be grouped with whom.

The efficiency of S play wins out in the long run if the grouping dynamics produces a concentration of S play and then continues to produce it—just as with C play in the Dilemma. For example, consider the grouping dynamics in which there is almost no movement of offspring. It is obvious that in groups to the right and left of the zone of contrary forces, the S and R fractions respectively will increase; and it can be shown that in fact the population will be pushed towards a state in which every group is a monomorphism (Myerson 1991), all-S or all-R. All the high S groups are fitter than all the high R groups, so all the R groups die out.

Hi-Lo is a limiting case of Stag Hunt in which the off-diagonal payoffs get very small. This case makes it possible to understand—what at first blush might surprise—that even in Hi-Lo all this remains true: for low values of α (the fraction in a group that has the A trait), individual selection works against the A trait; for high values, in favour of it; and the efficiency of A wins out in the end if, but only if, the grouping dynamics produces sufficient concentration of A-play.[15]

Stag Hunt and Hi-Lo exemplify coordination settings in which effective group behaviour involves no organization. Loosely, 'organization' is a system of information transfer within a group whose function is to produce combinations of actions that are efficient for some goal. [Organizations would have been the subject of chapter VII. The editors discuss them in section 5 of the conclusion.] Different situations demand different *organizational forms* (Campbell 1994). In the problem of calling a catch in chapter 1, we get a better outcome from optimal play in an organization that includes a caller than from optimal play with no organization. Game theoretically, the former situation is one with an extra option for each player. This is the option of playing her part in the organization. This extra option is such that all choosing it produces a best combination of behaviours better for the group than the best combination in the no-organization case. Likewise, in a situation in which there is a predator threat, there is a need for scanners, who then communicate to the others the presence of a predator. Just as, if the only traits in contention are the nonorganizational behaviours R and S (giving interactions of Stag Hunt form), group selection will favour groups in which there is much execution of the best combination of nonorganizational behaviours; so if there are also groups that practice organized interactions which raise group fitness above the maximum unorganized level, group selection will favour these. Groups with better *modes of organization* would do better.[16]

Because group selection works for optimal organizational form, and because the lifeblood of organization is information distribution, Wilson (1997) would expect humans to evolve to be not only independent decision-makers but also 'fully integrated networks'. The evolved individual would have an innate capacity to form '*part of* a group-level cognitive structure in which . . . tasks . . . are distributed' (p. 358). The italics are mine; when I discuss 'organizational man' [which would have occurred in chapter VIII], I will have occasion to return to Wilson's arresting suggestion.

5. The Proximate Mechanism

5.1 Existing Proposals

I come now to the production question: given that behaviours that make for good group performance have evolved in man, what is it that goes on in an individual at the time that produces this behaviour? The behaviour must be mediated by neural, biochemical and psychological processes and states, including conscious ones such as emotions and ways of reasoning. These processes are called 'proximate mechanisms'. The most proximate processes in the chain are what psychology concerns itself with when it seeks a theory of group behaviour. These most proximate processes are also the stuff of game theory, for game theory seeks psychological explanations of dispositions to choose in group contexts.[17]

Sober and Wilson point out that there are many conceivable proximate mechanisms for a given behavioural trait, so we need criteria for selecting among them. We may apply general theory-selection principles such as parsimony: the complexity of the hypothesis should be low relative to its explanatory power.[18] But there is another high-level criterion for proposed mechanisms M for a behavioural trait T. It is the test posed by the mechanism selection question: would M, a mechanism for the production of a trait T that could have been selected, itself have been selected? This criterion generates a number of more specific criteria, corresponding to properties which M needs to have to meet this criterion—availability, reliability and energy efficency.

In assessing the adaptiveness of proximate mechanisms, we need to pay attention to the fact of ludic diversity, which means that individuals need an array of behaviours for a range of situations with varying payoff structures. To simplify, suppose the ludic ecology consists of just two types of game, *n*-player Public Good Games à la Sober-Wilson, and *n*-player Stag Hunts. Call these game types G_1 and G_2. Suppose we have reason to believe—reason given by answering the behaviour selection question—that people evolved to have trait T_1 for G_1 and trait T_2 for G_2. A theory which postulates a single psychological mechanism M that subserves both T_1 and T_2 would, other things being equal, score better in terms of parsimony than one that postulates a different mechanism for each trait. Such a theory would also tend to score well on the selectability criteria of Sober and Wilson, since running a mechanism bears a fitness cost, and there are 'economies of scope' for a single mechanism that can support a range of traits. This is the more likely, as interaction problems of different types, G_1 and G_2, usually present themselves at different times. Suppose, simplifying again, that any proximate mechanism bears the same cost. If M supports appropriate behaviour in both G_1 and G_2, their nonsimultaneity means an M-possessor can behave adaptively in both for the fitness price of one mechanism.[19]

In a proximate mechanism for behavioural traits for interactive situations, we need a bundle of propensities—payoff-determining ones, epistemic ones and practical reasoning ones—which jointly yield the target behaviour. We must be careful not to lose sight of cognitive aspects of the problem faced by the reasoning agent. In particular, however we model the interaction itself, or the agent's motivations, because the agent faces a *game* the theory must also endow her with heuristics for resolving the special problems of strategic reasoning that bedevil games.

In the last few years several proximate mechanisms have been suggested for the disposition to play C in a Prisoner's Dilemma, which may or may not be repeated. Cosmides and Tooby (1989) and others suggest far-sighted egoism plus individual recognition plus memory for the trait of Tit-for-Tat in the repeated fitness Dilemma with low discount rates. Tit-for-Tat had been shown to be biologically adaptive through individual-level selection by Trivers and others some years earlier. Cosmides and Tooby pay careful attention to the bundle-of-propensities problem. Since implementing a strategy of the Tit-for-Tat type requires the agent to keep track of what happened on previous encounters with the current interactant, the proximate mechanism for the cooperative behaviour being explained must include person-recognition and outcome-memory abilities as parts of it, and these too are postulated.[20]

Proponents of the reciprocal altruism/long-term egoism theory argue that when scientists observe actions which they take to be instances of C in unrepeated Prisoner's Dilemmas, their model is misspecified, and the game the protagonists are really playing—or at least take themselves to be playing—is a repeated Dilemma. The far-sighted egoism theory doesn't work without this misspecification claim. In psychology and economics the claim that subjects in laboratory enactments of Prisoner's Dilemmas play, through habit or misperception, as they would 'out there', where it is typically one in a sequence, is not fully resolved, but the weight of evidence is against this 'assimilation' hypothesis.[21] The disagreement over this issue is a salutary reminder that unproblematic observational facts are not always easy to come by. It is not quite clear what behaviours we do observe; according to some, cooperation is not a fact to be explained, because it is not a fact.

Kitcher (1993) and Sober and Wilson (1998) propose that humans have evolved to be (psychological) altruists, that is, to be motivated to bring it about that others' preferences are satisfied. It is argued that the altruism hypothesis does better than the far-sighted egoism hypothesis by the 'criteria for choosing among mechanisms' hypothesis: availability, reliability and energetic efficiency. Altruism works by generating, in an interaction which is of n-player Prisoner's Dilemma form in fitness, one that is of quite another form in psychological payoffs or utilities. In the most favourable case for the theory, the utility matrix makes C dominant.[22] This case is favourable because no reasoning powers need to be assumed beyond those required for choosing a dominant strategy, and this is the most rudimentary reasoning ability imaginable for interactive decisions.

Gintis (2000) proposes 'strong reciprocity' as a proximate mechanism for the trait C-and-police (that is, play C and monitor others, punishing D and/or rewarding C) in repeated fitness Dilemmas. In the formulation of Falk and Fischbacher (1999), a strong reciprocator is someone who, if she thinks someone else is being 'kind' to her, gets utility from repaying him, and if she thinks he is being unkind, gets utility from harming him (so that she may have a desire to benefit, or harm, someone else even at a cost to herself). To be 'kind' here is to act with an intention to benefit. This psychological mechanism has been suggested not in an evolutionary context but simply as a hypothesis to explain behaviour in laboratory experiments. Gintis puts it to evolutionary use, first showing that a C-and-police behavioural trait could be biologically selected at the group level, then suggesting that strong reciprocity would support such a trait.[23]

Other proximate mechanisms that have been suggested for C involve social norms. A human social group may be characterized by social norms—rules that govern behaviour, learned, and generally accepted, in the group. Norms can be effective in at least two ways. They may be accompanied by a system of rewards and punishments for keeping to and violating them.[24] But also, it is said, people 'internalize' social norms, by which is meant that they come to have a preference for the actions picked out by the norms because they seem to be right in themselves, rather than because they are known to be socially approved. The internalized norm mechanism includes affective processes such as a tendency to feel shame at failing to do the acts in question, especially when the failure is observed by others.

Norms have the feature that they can both spread fast through a group of people living together and can also be handed down to the next generation like genes. If the norm is for a trait that is group-adaptive, such as a C trait, both these processes, of intra- and intergenerational transmission, amplify the between-group selection pressure for the trait. The former does so because it increases the correlation between the traits of different members of a group; the latter does so because it increases the assortativeness of the regrouping process (Boyd and Richerson 1985; Sober and Wilson 1998).

There appear to have been no proposals for proximate mechanisms for cooperative choices in explicitly modelled types of interaction not belonging to the Dilemma family, such as Stag Hunts and Chickens.

These various proposals are important advances, not least because they respect the vital distinction between fitness and utility and, more generally, the behaviour selection question and the production question. Sober and Wilson also explicitly address the mechanism selection question, and argue that altruism meets the criteria better than far-sighted egoism. But none of these proposals responds to the need for a *holistic* theory of the proximate mechanisms producing successful groups, one which explains psychologically our human ability to function well in a ludic ecology which has diversity. In particular, they are all Dilemma-directed theories, seeking mechanisms for traits that produce C in some interaction or other of the Dilemma family. They do nothing to explain, psychologically, cooperative behaviour in common-interest interactions, or participation in organized interactions.

To be specific, altruism cuts little ice in any pure coordination problem or any Stag Hunt.[25] Not only does this mean that we are left without an explanation of prosocial behaviour in non-Dilemma problems that are *'problèmes trouvés'*; but the hypothesis itself implies that when people confront Dilemmas these get payoff-transformed into non-Dilemma problems, as we have just seen, so the explanation is incomplete in two places.

5.2 Group Selection and Group Identification

I suggest that group identification is the key proximate mechanism in sustaining cooperative behaviour in man. More fully, I conjecture this: dispositions to cooperate in a range of types of game have evolved in man, group identification has evolved in man, and group identification is the key proximate mechanism for the former. The main virtue of this hypothesis over that of altruism and other contenders is that group identity is a more powerful explanans of the diversity of cooperative behaviours we see. Group identity implies affective attitudes which are behaviourally equivalent to altruism in Dilemmas, and it can explain what altruism cannot, notably human success in common-interest encounters.

Long ago, Durkheim (1893) surmised that early human life was characterized by collective self-identity: 'The primitive state of the group is that of an undifferentiated crowd of human beings, none of whom is pushed to individuate himself from the others, since the acquisition and distribution of goods is carried out in common'.[26*]

Caporael (1995) suggests that early life was characterized by participation in a number of groups, of different orders, which she calls 'core configurations' (dyads, face-to-face groups, demes and macrobands). She elaborates on Durkeim's speculative hypothesis that humans identified with their groups. She hypothesizes that because their memberships were multiple, so were their self-identities. According to

Caporael, identifying with a core configuration was—and is—part of the proximate mechanism that maintains it, because it facilitates working successfully in a group. She also holds that a capacity for identifying with these groups helps explain the continual 'reassembly' that characterizes them. We may conjecture that the way this capacity facilitates reassembly is that entification facilitates re-entification (once a gestalt, always a gestalt).[27]

Caporael thinks that early humans switched their activities from group to group with situational and other conditions, and that this plasticity is mirrored by the way we, today, switch group identities with shifts in shared outcomes, interdependencies, gestalt boundaries and other conditions. She points out that the proximate mechanism for group activity evolved for small groups, long before 40,000 years ago, but also clearly worked for the larger groups, for example for those we lived in after the discovery of agriculture, so it must be activated by factors which do *not* depend on group size. Once again group identification fills the bill, since we seem able to identify with, and lay down our lives for, groups of still higher orders of magnitude, such as classes and nations. She takes these parallels between the group-formation conditions and group-identification conditions to be evidence that identification is a proximate mechanism for formation.

I come now to my own evolutionary argument for the hypothesis that group identification is the fundamental evolved proximate mechanism for collaboration in man.

First, there is reason to think that man has evolved to have a repertoire of cooperative behaviours geared to different types of situations with scope for cooperation. (The cooperation may be with all others in the interactive situation, or only with a subset.) The situations include both Dilemma-type situations and common-interest situations. Call the hypothesis that people are endowed with such a range of cooperative, situation-dependent behaviours the *cooperative repertoire hypothesis*.

Part of the support for the cooperative repertoire hypothesis is ordinary observational evidence, for what it's worth. As we saw earlier, you cannot just read off from observations of behaviour whether or not people have a given trait, even when it is behavioural rather than psychological. Interpretation is involved. Nevertheless, the evidence is about as strong as it ever is that people often (though by no means always) engage in cooperative behaviour in both Dilemma-type games and common-interest games such as Stag Hunts and Hi-Los.

But there is another reason to believe the repertoire hypothesis, which is a fruit of this chapter's excursion into evolutionary theory. This is that natural selection at the group level is possible. If it did happen it would, as is now well known, have favoured C traits for

Dilemma-type interactions. But it would *also* have favoured S traits for Stag Hunts, A traits for Hi-Los, and appropriate cooperating traits for other types of common-interest interactions (for example, traits for resolving the coordination problems involved in achieving efficiency in Chickens and Battles of the Sexes).[28*] If in the past there was selection for C at the group level, the same between-group forces would have selected for group well-functioning in common interest interactions.[29]

The right production question to ask, therefore, is what is the mechanism for this *repertoire* of cooperative traits. To answer the production question, this repertoire must be comprehensive—it must prime all the components in the repertoire in the appropriate situations. Suppose that traits have evolved for S in Stag Hunts, A in Hi-Los, and C in Prisoner's Dilemmas. Then the proximate mechanism must produce S in Stag Hunts, A in Hi-Los, and C in Prisoner's Dilemmas.

Furthermore, the mechanism should not produce unwanted byproducts. M does not answer the production question if, in addition to producing the repertoire, it is bound also to produce further traits which would not have been adaptive—for example, if the trait is to behave self-sacrifically towards comembers of a group, the proposed mechanism should not be one that would also produce self-sacrifical behaviour towards nonmembers. Generated behaviour must be targeted. Finally, to satisfy the criteria for a good answer to the mechanism selection question, it must be as simple and compact as possible—ideally we would like a single multipurpose mechanism.

As we have seen in the last subsection, altruism by itself fails the comprehensiveness condition. Some versions of altruism also fail the targetedness condition, because they spread the benefits of sympathy to all and sundry.[30]

On the other hand, group identification is, I suggest, comprehensive, targeted and multipurpose. Showing this claim depends on understanding the details of how group identity gives rise to C, S, A and so on. Some pieces of the picture we already have: in the last chapter we saw empirical evidence that group identity increases cooperation, and that it transforms egoistic motivations into ones directed to group goals. This transforming feature means that the group identity theory partially subsumes the altruism theory. For although wishing for the good of the group is not the same thing as wishing for the good of other members of the group, it comes close to it in the case in which the group good is Paretian: both altruism and group identification are conative mechanisms which affect what an individual wants, and in the Paretian case both induce the same transformation of the private ordering of outcomes. Group identity solves Dilemmas as well as altruism does, because it generates altruism in the appropriate groups. But, unlike undiscriminating altruism, it generates it only in groups which have a common evolutionary interest.

Last but by no means least, it also produces cooperation in common-interest games. In the following chapters I will say how this may be. With this promissory note, I end my excursion into evolutionary theory. I conclude that what we have found there was worth the detour. Thinking about the evolution of group behaviour supports the thesis that group identification is at the heart of cooperative human behaviour in cooperative games of both adverse-incentive and common-interest kinds. Such thinking supports this thesis because it shows us that selection processes would favour a mechanism that subserves cooperation in both kinds of interaction, and because group identification is such a mechanism.

6. Stocktaking

In this chapter I have surveyed a body of evolutionary theory which starts out from evidence that man evolved by natural selection to function well in groups, and argues that this could have happened by selection at the group level. This literature thus offers a group-selectionist answer to the 'behaviour selection question' of why and how certain innate behavioural propensities—here those involved in well-functioning in groups—might have evolved in the wild. But if indeed man evolved to be (sometimes) a good group member, it becomes worthwhile to ask the 'production question' of what the 'proximate mechanism' is—what psychological processes produce the behaviours in individual human beings. In seeking an answer to this question, we must return to the subject of natural selection and ask the 'mechanism selection question' of how and why one rather than another proximate mechanism would have been selected. I have argued that the process known in social psychology as group identification is the, or at least a, key proximate mechanism in producing well-functioning in groups. There are two things in particular that argue for this hypothesis. The first is that group identification is thought to be activated in a range of group settings, and these match the range of group participations we find in the activities of early man. The second is that it potentially explains cooperation in common-interest interactions as well as in Dilemma-type interactions. The hypothesis therefore satisfies, better than alternatives such as altruism, the main criteria for the selectibility of a proximate mechanism articulated by Sober and Wilson.

Notes

1. When we have answered all three questions we have given an evolutionary explanation of the to-be-explained behavioural trait T: T is favoured by natural selection, and produced by a mechanism M that is favoured by natural selection. An explanation of this kind sometimes needs to be supplemented by an appeal to *domain-transferability*. If T is a disposition to act with others in the defence of territory at the risk of personal injury, supplying a story for its natural selection and a suitable M almost but not quite explains why footballers play with more 'commitment' in home games; but it does explain it if we think that the evolved mechanism *M* operates in response to features of the original situation which don't change as we go from territory to stadium and from axe injuries to boot injuries.

2. Evolution is strewn with examples of adaptations rendered obsolete by changing ecology, such as the excess brain capacity of gibbons and orangutans (Dunbar 1993).

3. Different people stress the roles in the success of *Homo sapiens* (500,000–100,000 B.C.E.) of groups of different levels and different makeups. For example, the evolutionary economist Gintis (2000) writes: 'The stunning success of *Homo sapiens* is based on the ability of its members to form societies consisting of large numbers of biologically unrelated cooperating individuals.' That the cooperators need not be kin argues against the theory that cooperativeness is largely the product of 'kin selection', which explains the evolution of a trait T as a result of continued interactions of groups of kin. Another factor suggesting that kin selection is not the most plausible story is that there is little match between kin groups and cooperating groups and we often find sharp internal conflict in groups of kin (Ross 1986). Most social scientists think that today's ethnic identities, and hostilities, are socially transmitted (Brewer and Miller 1996).

4. The mean group size in the Tai chimpanzees observed by the Boeschs was 4.1. If there are enough in the team, they collaborate by dispersing below the prey (colobus in the treetops), out of each other's sight but focussing on the same prey, wait for the opportunity to act and reunite to corner the prey.

5. After the allocation, a subgame ensues in which individuals' options include, for nonowners, theft, taking hold and recovery (of food dropped by another) and, for owners, passive sharing (not reacting to someone's taking hold) and proffering.

6. I suspect a strange and somehow beautiful reflexivity: why do we label certain behaviours b_1, b_2, b_3 'achievements' for the group, ways of 'functioning well'? Because we have an intuitive notion of achievement for the group. Where does it come from? From the fact that it was selected for. Why was it selected for? Because it was part of the proximate mechanism that drove b_1, b_2, b_3, and b_1, b_2, b_3 favoured those who practised them, at the group level.

7. The core of the notion of cooperating is acting together in such a way that an outcome results that is good in terms of some objective (thus it is objective-relative). In the context of intentional reasoned action, which is what most of this book is about, we may want to gloss this in various ways. For one thing, we may require that the individuals intend to produce an outcome of this kind.

For another, some may want the objective to be a common interest of the individuals. For another, we may want to be specific about what would happen otherwise, for example tying this to a theory of independent action. I shall come back to these refinements in chapter 4.

8. Neither do game theorists: although game theory studies games of very varied structures, there has been almost no interest in what happens in the game of life, or even short sequences of diverse interactions. The basic reason for this is that game theorists are interested in a form of reasoning that allows a player to adapt her actions perfectly to the current problem. If players have game-playing capacities that work well for some game types but not others, the mix matters. In fact, this *is* the case for the capacities expressed in standard solution concepts, for these do not succeed in giving perfectly adapted play.

9. Dunbar and Nowak have both concurred with this view in conversation.

10. The model supposes that there is a large mixed population of C and D traiters, and that groups arise by random matching. Let the Prisoner's Dilemma payoffs be 3, 2, 1, 0. Let the fraction of C-ers in the population be γ. Consider an arbitrary C-er. It has probabilities γ, $1 - \gamma$ of interacting with a C-er or a D-er, and so an expected fitness gain per encounter of $2\gamma + 0 = 2\gamma$. An arbitrary D-er, meanwhile, has an expected fitness gain of $3\gamma + 1(1-\gamma) = 2\gamma + 1$, which is greater, for all γ.

11. A particular kind of correlated repartitioning is kin grouping: offspring tend to stay together, with the result that (assuming traits are inherited), T-traiters tend to interact with T-traiters.

12. The Price equation concerns a population containing an arbitrary number of types. There are several groups $g = 1,2,\ldots$. We are interested in how the population fraction of a particular type, say A, changes from one generation to the next. Write α_g for the A-fraction in group g, $\bar{\alpha}$ for the population A-fraction, and let Δ denote change of value in a variable from generation t to $t + 1$. (Thus $\Delta_{\alpha g}$ is $\alpha_g(t + 1) - \alpha_g(t)$.) Write f_g for the fitness of an average member of group g, and $\bar{f}$ for the fitness of an average member of the population. The Price equation, in its most perspicuous form, is

$$\bar{f}\Delta\bar{\alpha} = \beta\ \mathrm{var}(\alpha_g) + \Delta\alpha_g f_g$$

The left-hand side measures the spread of A. $\Delta\bar{\alpha}$ is the increase in the population A-fraction from t to $t + 1$; it is premultiplied by the current overall fitness level of the whole population, but for present purposes we can think of this as a constant. On the right-hand side, the coefficient β is the regression coefficient of f_g on α_g, which equals $\mathrm{covar}(f_g,\alpha_g)/\mathrm{var}(\alpha_g)$ and measures the rate at which group fitness f_g rises with group A-fraction α_g across groups. High β means that having a lot of A has a big positive effect on group fitness, as we might expect in a Dilemma. High β will clearly help A to be selected at group level, but only, as we have seen, if A is concentrated in some groups. The extent to which it is, is measured in the Price equation by $\mathrm{var}(\alpha_g)$, the intergroup variance of the A-fraction. The whole first term, $\beta\ \mathrm{var}(\alpha_g)$, gives the *group selection effect* on $\bar{\alpha}$. The second term on the right-hand side is the individual or *intra-group* selection effect: $\Delta\alpha_g$ is the increase in the A-fraction *within* group g (which is negative in all groups in a Dilemma world); it gets multiplied by f_g

because if g is fit it bulks larger in the population in generation $t + 1$, and so what is happening to A in it has more weight in determining $\Delta\bar{\alpha}$.

13*. This mechanism works for games such as Stag Hunt, in which the collectively optimal strategy is the best reply to itself. In games such as Prisoner's Dilemma, in which the collectively optimal strategy is not a best reply to itself, spatially isolated groups of cooperators are vulnerable to invasion by noncooperative 'mutants'. In such games, the survival of cooperative behaviour may require a continuing process of assortative regrouping, the effect of which is to expel noncooperators from associations of cooperators.

14*. In MB's manuscript, figure 3.2 appeared only in sketch form and was not self-explanatory. The editors have redrawn the diagram and revised the text to correspond with the new diagram and to explain the derivation of the results. The substance of MB's argument has not been changed.

15. Consider this time a cultural selection process. Let A and B be two different conventions, one of which, A, has some intrinsic advantage. For example, A is the male teenager practice of blading and B that of boarding, and B produces more serious injuries than A. I stylize the facts. Each day an individual gets paired with another member of his social group at random. A's who find themselves with other A's have a good experience; similarly for B's. Injuries are bad experiences. From time to time (say at the end of each school term) people who have had good experiences all in all talk up their own practice to their friends, who sometimes switch. Each new term social groups are re-formed, perhaps because classes are. In any term, in groups with mostly bladers, individual selection favours blading; in ones with mostly boarders, it favours boarding. The better safety prospects of blading leads blading to win out if this process runs for years and years if, but only if, the termly clique-formation process makes bladers hang out with bladers to a sufficient degree.

16. Campbell (1994) is interested in explaining, in particular, the organizational form of the modern firm. He suggests that the modes of organization we see are those which have come through because they are selected at group level against selection pressure at lower level. But he posits a three-level structure: high-level configurations of middle-level groups he calls 'face-to-face' groups, which are groups of individuals. In his model, face-to-face groups cooperate internally and so get selected at the middle level. But they practice 'clique selfishness' and so endanger the high-level units to which they belong. The informal groups at the Hawthorne plant were of this kind. The high-level units will therefore be selected only if their organizational form includes norms and sanctions that restrain the selfish cliques.

17. Psychology, game theory and evolutionary theory have different special concerns. For example, the rationalistic tradition in game theory restricts the proximate mechanism to being one that is rational. Cognitive psychology restricts it to being one that is consistent with known cognitive capacities. In evolutionary theory the production question is accompanied by another, the mechanism selection question, which restricts the proximate mechanism to being one that is likely to have evolved.

18. A virtue claimed for rational choice theory and rationalistic game theory is that a single mechanism (expected utility maximization together with rational

principles that mandate Nash equilibrium) predicts behaviour in a very wide range of decision situations. It is true that the axiom basis is compact, and this makes for parsimony. But unfortunately the theory does not do very well as a predictor of observed behaviour. For example, we observe robust violations of the predictions of expected utility theory, and Nash equilibrium game theory fails to predict A in Hi-Lo.

19. Sober and Wilson (1998, p. 307) make a similar point in an example comparing the adaptiveness of magnetosomes and oxygen detectors as proximate mechanisms in a marine bacterium in navigating itself to avoid oxygen. They say: 'perhaps the bacterium already uses magnetosomes for other tasks. If so, it might be more efficient for the organism to use this device for the additional purpose of navigation, rather than add a second piece of machnery to achive this end. The adaptations . . . require energy to constrict and maintain. A prediction concerning what proximate mechanism will evolve must take account of *energetic efficiency*.'.

20. Not even Cosmides and Tooby's bundle is a complete outfit. We must also model agents' reasoning procedures. In the present case, people must be ascribed some mode of reasoning about repeated games which delivers Tit-for-Tat as a solution to the problem of strategy choice. [MB planned to write something here on laboratory and other evidence on whether people play Tit-for-Tat and, if so, how they are reasoning.]

21*. MB intended to write a note here, presumably citing the relevant empirical evidence.

22. It is assumed that individuals have basic desires (such as satisfying hunger) which correspond to fitness consequences, so that the matrix of primary utility payoffs is the same as the fitness matrix. [Consider the Prisoner's Dilemma, shown in figure 3.a. If the players are symmetrically altruistic, their utilities are transformed to $U_1 = \lambda_{u1} + (1 - \lambda)u_2$ and $U_2 = (1 - \lambda)u_1 + \lambda_{u2}$. If the payoff transformation is golden rule altruism (λ =0.5), and if the original game has the parameters a = 5, b = 4, c = 3, d = 2, we get a game with transformed payoffs a' = 3.5, b' = 4, c' = 3, d' = 3.5. This explains the choice of C, by dominance. But if the original game has the parameters a = 5, b = 4, c = 3, d = 0, golden rule altruism implies a' = 2.5, b' = 4, c' = 3, d' = 2.5, a Hi-Lo, while a value of λ between 0.6 and 0.8 gives a game of yet another form, a Stag Hunt.] For the second Dilemma, therefore, a proximate mechanism must include reasoning procedures that resolve problematic games.

		Player 2: *cooperate*	Player 2: *defect*
Player 1	*cooperate*	*b, b*	*d, a*
	defect	*a, d*	*c, c*

Figure 3.a. The Prisoner's Dilemma

23. In this theory the behavioural trait and the proximate mechansim are broadly isomorphic, because the mechanism is in effect just a desire to behave

The choice mechanism in the trainer case is an example of what I shall call 'simple direction'. It will be useful to define this in a little more detail. Take any simple coordination context (T,O,U). Let o^* be the profile o that is best in terms of U.[3] An $(n + 1)$st agent, the *director*, works out o^*, identifies the agent in control of the ith component action, and tells this agent i to do her component o_i^*; agent i executes her instruction. Note that the director goes through a three-stage process, first computing, then identifying, then instructing. If everyone in T is governed by simple direction, the profile o^* is executed and the goal expressed in U is achieved (in the sense that the maximum feasible value of U given the options is attained).

Team reasoning is do-it-yourself direction. In a simple coordination context (T,O,U), an agent i is said to *team-reason* when she decides what to do in the following way. She first computes o^*. She next locates the element o_i^*. Last, she reasons that she should perform o_i^* because this is the component of the best profile that is under her own control. Usually it's trivial to locate o_i^* once you have computed o^*, but it need not be.[4] If everyone in the set team-reasons then, once again, the profile o^* is executed and U is maximized.

It is helpful to think of the first step of team reasoning by i as i asking herself, 'What would the director have me do?'—that is, as simulating a director. This is helpful because it enables us to separate two aspects of what a team reasoner does: one (step one) is to reason *at the group level*—to engage in profile-based reasoning—and the other (step two) is to reason *as an individual* that because the ith component of the output of stage one is what it is, that's what she should do.

Team reasoning seems like a good thing. You don't need a director. But we must consider whether the team-reasoning mechanism for T is likely to be feasible. For it to be, all the agents in T must have the means to compute the best profile for U, o^*. In particular, all must know U, the set O of feasible profiles, and the values of all the profiles in O: in brief, all must know the payoff structure.[5] This is not a negligible condition; there are many simple coordination contexts in which agents are not aware of, for example, all the options of other agents. On the other hand, there are very many in which everyone is aware of the whole payoff structure, at least to an approximation.

In a simple coordination context, if the decisions of the members of T are reached through any mechanism that ensures that their common goal U is achieved, I shall call the mechanism a *team mechanism* and T a *team*. The reason for this terminology is that an important part of the ordinary notion of a team is a set of agents effectively coordinated to achieve a single goal. The mechanism I have just defined as 'team reasoning' I label thus because the members of T are brought to do their parts in the best profile for U by pure reasoning.

Figure 4.1. Offside trap

Football example: Offside trap. Figure 4.1 shows a position in a football game between Chelsea (grey) and Liverpool (black). Chelsea are on the attack. Number 9 has the ball for Chelsea. The arrows emanating from 9 and 10 show the intentions in the minds of these players; together they make up a profile of intentions. Number 10 intends to move into space past the defending player 3 and shoot for goal if the ball comes to him; 9 intends to slot the ball between defenders 2 and 3 into 10's path. The diagram also shows two profiles of options for the Liverpool players, and these are the only ones in their minds. In the first option (solid arrows), 2 tackles 9, 3 blocks 10's path, and 4 runs back to the near goalpost to block 10's shot in case the Chelsea manoeuvre works. The goalkeeper 1 stays on his toes. In the second (dashed arrows), number 1 does likewise and all three defenders 2, 3 and 4 run forward. If they are all farther from the goal than 10 at the moment when 9 passes to 10, 10 will be 'offside', Liverpool will win a free kick and the attack will be thwarted. The dashed profile is better for Liverpool than the solid profile: it is virtually certain to work (the referee is a top professional, and the only serious danger is an elementary refereeing failure), while 9 and 10 might evade 2 and 3 respectively, and then 10 would be in a good position to shoot.

Call the solid profile c and the dashed profile d. There is a simple coordination context (T,O,U) in which T consists of the four Liverpool players 1, 2, 3, 4: O consists of c and d; and U is -1 times Chelsea's expected score. In this example we might expect all the members of A to share psychologically the aim U. If so, we can add Offside Trap to the examples of Hi-Lo in chapter 1. But this is as may be.

If (for whatever reason) each Liverpool player team-reasons for U, each selects d, so 1 decides on d_1 (to stay on his toes), 2 on d_2 (to run forward), 3 on d_3 (to run forward) and 4 on d_4 (to run forward). This is not the only way that T could arrive at the efficient profile of options d. Suppose Houlier, the Liverpool manager, were able to transmit high-speed messages to his players, by telepathy or some futuristic IT. Houlier could see that d was better than c, and could instruct 1, 2, 3 and 4 to do d_1, d_2, d_3, d_4 respectively. If the players obey the manager's instructions, d results.

Whenever the taskforce goal U is Paretian, team reasoning is 'mode-P reasoning' in the language of chapter 1, section 8.2, satisfying both the defining principles of mode-P reasoning, F1 and F2. A team reasoner for U ranks all act-profiles using a Paretian criterion, just as F1 specifies. And she takes herself to have a reason to enact her component in the highest-ranked act-profile (since team reasoning involves deciding to do o_i^* because this is the component of the best profile that is under her own control), just as F2 specifies.

There are many conceivable team mechanisms apart from simple direction and team reasoning; they differ in the way in which computation is distributed and the pattern of message sending. For example, one agent might compute o^* and send instructions to the others. With the exception of team reasoning, these mechanisms involve the communication of information. If they do I shall call them *modes of organization* or *protocols*. [MB intended to discuss the communication of information in chapter VIII. The editors take up this topic in section 5 of the conclusion.]

There are also plenty of mechanisms that are not team mechanisms in our sense (that they ensure that U is maximized). The process in which each agent picks one of her options at random, like a Stahl-Wilson level 0 player, determines choices for the set of agents which fail in general to maximize U. The processes described in game theory are more difficult to classify: according to game theory each player deliberates in accordance with individual rationality, but game theory fails to tell us the outcome of these deliberations; in particular, in simple coordination contexts the standard position—that the deliberations lead to profiles that are Nash equilibria—does not tell us enough to say whether they constitute team mechanisms, since there is more than one equilibrium, of which some do and some don't maximize U.

A mechanism is a *general process*. The idea (which I here leave only roughly stated) is of a causal process which determines (wholly or partly) what the agents do in *any* simple coordination context. It will be seen that all the examples I have mentioned are of this kind; contrast a mechanism that applies, say, only in two-person cases, or only to matching games, or only in business affairs. In particular, team reasoning is this kind of thing. It applies to any simple coordination context whatsoever.[6] It is a mode of reasoning rather than an argument specific to a context.

Being a team mechanism seems to be a good thing. A team mechanism M has the outcome that the agents' choices make up o^*, the best profile in terms of U, which evaluates profiles in terms of a goal all the agents share. But this is not enough to endorse M. If it seems to be, that is because we are misled into forgetting about the events that precede the choices in o^*, including the actions that M requires the agents to take, such as thinking, and reading messages. These events may affect the agents' interests other than through the profiles that they lead the agents to choose. Before we endorse M we need to consider the costs of running M as well as the merits of the decisions to which it leads. This is Simon's (1979) famous point that agents and actions are to be evaluated not only as more or less 'substantively rational' but also as more or less 'procedurally rational'. [MB intended to take up this issue in a later section of this chapter, but had not yet written it at the time of his death.] But for the time being I put this matter to one side, ignoring questions about the procedural costs of team mechanisms in order to concentrate on other issues.

We now know what it would be for members of A to team-reason. We shall see soon that team reasoning is a powerful engine for producing decisions—it is determinate in a much wider range of cases than standard game-theoretical reasoning—and for producing good ones. But for all I have said so far, there is no obvious reason to think that people actually do team-reason. As it happens, I think that in the football case not only are the players very likely to have the aim *U*, but they are also quite likely to team-reason, and that this is true of members of taskforces in a class of cases of which this is exemplary. Later, I shall also present experimental evidence that people team-reason. But before we get to any of this I have to deal with an apparent flaw in team reasoning.

3. The Problem of Failure

3.1 Restricted Team Reasoning

Take any simple coordination context (*T,O,U*). If the behaviour of *T* is governed by team reasoning, the best profile o^* in terms of *U* is executed. But suppose that the mechanism fails to work in some agents; what then? In this section I address this problem—*local failure*—and show that team reasoning, properly formulated, emerges unscathed.

If one part of an optimal configuration of a set of variables is somehow blocked, the best values of the remainder may be affected. This is the basic theorem of the second best.[7] It is best for you and your partner to meet in Geneva, but if he is held up in Prague, it is best for you not to fly. For a more complex example, consider a three-person version of Stag Hunt with players P1, P2, and P3. Let a stag be worth 60 to the group, a rabbit 10, and an injury −10. Suppose that whoever hunts for rabbit gets one, and that if three hunt the stag they will get one for sure, that if two do they have a 2/3 chance, and that if one does alone he will fail and be injured. If three choose S, the group's payoff is 60; if two choose S, 50; if one does, 10; if none do, 30. Suppose that for some reason P3 chooses R; then the best-for-*U* profile for P1 and P2 is the 2-long subprofile (S,S). So in this case participating in the first-best profile o^* gives the right answer. But suppose that for some reason both P2 and P3 choose R; then the best-for-*U* profile for P1 is the 1-long subprofile (R), so participating in o^* gives the wrong answer.

In the version of team reasoning that I commend, if P1 and P2 have common knowledge that P3 will choose R, then team reasoning tells *P1 and P2* to choose S, because this is their part in the best-for-*U* profile for them given P3's non–team-reasoning choice. If P1 knows that P2 and P3 will choose R, then team reasoning, I claim, tells P1 to choose R, not S. In this version of team reasoning, the team reasoners may make up only a subset of the agent set, and team reasoning requires them to optimize *U* as best they can between them, doing without the non–team reasoners.

Formally, instead of a simple coordination context (T,O,U), we have a system that can be represented as (S,T,O,U,f), where now S is the set of n agents who have the possible profiles of options O on which U depends, and T is a subset of S. Writing $R = S - T$, f is the subprofile of o giving the choices of those in R, the *fixed subprofile*. R is called the *remainder*. I call such a situation a *restricted coordination context*. The possibilities for coordination are restricted by the fact that the choices of those in the remainder are fixed. An option profile is now $o = (v,f)$, where v is the subprofile of choices of those in T, the *variable subprofile*. Restricted coordination contexts generalize simple coordination contexts, which are the special case in which $S = T$.

In a restricted coordination context, team reasoning is defined as follows. Each agent i in T determines v^*, the maximizer of U over v given f, then identifies and executes v_i. This procedure, team reasoning in a restricted coordination context, I shall call *restricted team reasoning*. For restricted team reasoning to be feasible, all in T must know the payoff structure and the identities and choices of those in the remainder. Notice that it is only the members of T who are assumed to maximize U (as best they can). Since f could be anything, there is no presumption that the behaviour of those in R is in any way guided by U.

In a restricted coordination context, the mechanisms of interest are processes that (wholly or partly) determine the choices of the agents in T, but have no causal influence on the choices of the remainder. If M is a mechanism, I'll say that the members of the remainder *fail* under M and that the members of T *function* under M. It is easiest to assume that which agents fail and which agents function is the same under any mechanism (that T is independent of M), and I shall talk as if this is so. But in fact the model can accommodate the case in which M can affect T, and this is certainly possible in reality: for instance, some people might fail to team-reason because of their cognitive limitations, and others might fail under coercion because of their cussedness.

In a restricted coordination context I'll call a mechanism a *team mechanism* if it produces the best possible outcome in terms of U. What is a possible outcome is now restricted by the fact that R will perform f; otherwise, nothing changes. Once again, there are many alternative team mechanisms one can think of. In particular, restricted direction: a director computes v^* and instructs each i in T to perform her part v_i^*. This is what happens if Houlier on the touchline sees that Owen, the Liverpool striker, is limping. If Owen is limping he will be unable to carry out his instruction o_i^*,[8] so simple direction cannot operate on S. The team is down to ten men, and Owen is, for the time being, a passenger, constituting a one-agent R, and f is the one-place subprofile (limps).[9] As before, I call a set of agents whose behaviour is governed by a team mechanism a *team*. In a restricted coordination context, therefore, T is a team if its behaviour is produced by (for example) restricted team reasoning or restricted direction.

3.2 Cooperative Utilitarianism

An old proposal in ethics is that what it is right to do is to act so as to promote the total utility of the population. I shall write H (happiness) for this utility measure. This doctrine, utilitarianism, faces the difficulty that since total utility, H, depends on the actions of others as well as yourself, it is unclear what act of yours will best promote H. Different answers to this question are found in different versions of utilitarianism, notably act utilitarianism and rule utilitarianism. The difficulty is profoundly game theoretical. Suppose everyone wants to maximize total utility. Then the question becomes how to solve a coordination problem in which the common ranking happens to be given by total utility. The difficulty is confounded by the fact that it is an empirical question whether others also wish to maximize H and, if so, whether they follow the same version of utilitarianism.[10]

Regan (1980) distinguishes two intuitions in the writings of utilitarians confronting the first of these difficulties. One is that the correct theory of the right is one that's good for individuals to follow; the other that it is one that's good for all to follow. The first leads to the version called act utilitarianism and the latter to rule utilitarianism.[11] The former can be formulated in the following principle: in any decision problem, each agent i should choose that act which maximizes H given her situation. Her 'situation' includes all aspects of the state of the world she cannot influence, and so includes others' actions. Thus this principle, the central tenet of act utilitarianism, is a best-reply principle for H-maximizers.

The second intuition can be formulated thus: the correct theory has the property (C) that if all follow the theory in all decision problems then H is maximized. Regan calls property C the 'coordinated optimization principle'. To see more clearly what this is saying in purely decision-theoretic terms, let us strip away the ethical dimension. What we are left with is the class of coordination games with an arbitrary common maximand U. Let Θ be any theory of such games. The property C becomes the following property—say, C′—of Θ: if all choose in accordance with Θ then U is maximized. Thus a theory Θ of coordination games has property C′ only if it says that, if all conform to Θ, then all play their parts in the maximizer of U. This has a remarkable implication for game theory. The classical framework of game theory assumes both that there is a single form of rationality and that all players choose by it. It then seeks to pinpoint this form of rationality. Now suppose we have the second intuition, but we have it about coordination games in general—we think that a correct theory Θ has property C′. But (in the classical framework) all conform to Θ. Hence Θ must say that all play their parts in the maximizer of U. In particular, [if U is Paretian] the correct theory of Hi-Lo says that all play A. In short, an intuition in favour of C′ supports A-playing in Hi-Lo if we believe that all players are rational and there is one rationality.

Regan notes that property C doesn't get you very far, because if you think there are any nonfollowers it tells you nothing at all about what to do. He sets out to generalize C to remedy this deficiency. Regan's proposal is *cooperative utilitarianism*. Its key property, called 'adaptability' by Regan, is that those who do follow it maximize H given the behaviour of those who do not follow it. Cooperative utilitarianism requires an agent first to determine who the followers and nonfollowers are, and what the nonfollowers will do, then to play their parts in the combination of actions that maximizes H given these constraints. In our terms, the cooperative utilitarian engages in restricted team reasoning. Cooperative utilitarians form a *team* T in the restricted coordination context (S,T,O,H,f), where S is the whole population, O is the set of all possible complete profiles of options, and f is the subprofile chosen by the nonfollowers. If the entire population consisted of three stag hunters 1, 2, 3, and 1 and 2 were cooperative utilitarians who knew that 3 meant to hunt the rabbit, 1 and 2 would hunt the stag because they believed it right to do so.

Arguments for rule utilitarianism suggest that, in an important but special case of coordination, unrestricted coordination for H, a priori moral argument supports team reasoning; Regan's aruges that it supports restricted team reasoning in a more general context, restricted coordination for H. I'll return to all this when I consider the rationality of team reasoning. [This section was still unwritten when MB died.] But first we have to deal with another problem.

3.3 Circumspect Team Reasoning

Restricted coordination contexts are much more general than simple coordination contexts, and so restricted team reasoning is more general than simple team reasoning. But restricted team reasoning still has two serious limitations: first, for it to be feasible the reasoners must know who is and who isn't a team reasoner, while in practice there is very often great uncertainty about who is and is not governed by a choice mechanism. In moral decision-making, a utilitarian agent is typically quite uncertain who else is a utilitarian. In the case in which the mechanism is simple direction, the director may be unsure who has received her messages. In an army unit governed by some mechanism that aims for victory, it is always possible that some components may fail by death or desertion. Industrial workforces are subject to sickness and absenteeism. Call the neglect of this uncertainty about the remainder the *unknown remainder* problem for restricted team reasoning.

Second, the mechanism only applies when the choices of the remainder, f, are 'given', and known to the team reasoners; but this may not be so, and instead there may be a game between the team reasoners and the remainder. I shall return to this problem, which I call the *strategic remainder* problem. But first let us turn our attention to the unknown remainder problem.

There is no difficulty of principle in repairing this limitation of the restricted coordination context model. All we need do is to make failure (membership of the remainder) random. For simplicity I model the uncertainty about the remainder thus. Let M be a mechanism. I assume that every agent functions under M with probability ω, and this is known to all. Each of the n agents i in S has an option f_i which she does if she turns out to be in the remainder; f_i is called her *default choice*. Thus the n-tuple f of default options has the same dimension, n, as o. Instead of a restricted coordination context (S,T,O,U,f), we now have a system that can be represented as (S,ω,O,U,f) , where ω is the probability of functioning and f is the profile of default options. I'll still denote by T the subset of S that functions, but now instead of being fixed this is random. If S is the set of three hunters and $\omega = 0.8$, T will be {1,2,3} with probability $(0.8)^3 = 0.512$, {1,2} with probability $(0.8)^2(0.2) = 0.128$, and so on. I call such a situation an *unreliable coordination context*. I call it thus because agents cannot be relied upon to function under choice mechanisms. This is not quite a generalization of restricted coordination contexts, but it could easily be made into one by letting the probability of functioning be different for different agents.

To generalize the notions of team mechanism and team to unreliable contexts, we need the idea of the profile that gets enacted *if* all agents function under a mechanism. Call this the *protocol* delivered by the mechanism. The protocol is, roughly, what everyone is supposed to do, what everyone does if the mechanism functions without any failures. But because there may well be failures, the protocol of a mechanism may not get enacted, some agents not playing their parts but doing their default actions instead. For this reason the best protocol to have is not in general the first-best profile o^*. In judging mechanisms we must take account of the states of the world in which there are failures, with their associated probabilities. How? Put it this way: if we are choosing a mechanism, we want one that delivers the protocol that maximizes the *expected value* of U

		Player 2	
		C	D
Player 1	C	2, 2	−3, 3
	D	3, −3	1, 1

Figure 4.2. A Prisoner's Dilemma

given the probabilities of failure states. Label this protocol o^{**}. Formally, o^{**} is the o that maximizes the expected value of U given that, for each i, i will choose o_i with probability ω and f_i with probability $1 - \omega$.

Consider for example the Prisoner's Dilemma in figure 4.2. Suppose we are interested in mechanisms that maximize U, where U is the sum of the individual payoffs, and suppose that anyone in the remainder chooses D, for example because she reasons game theoretically. It's obvious [from the fact that U is always higher when both players make the same choice than when their choices are different] o^{**} must be either (C,C) or (D,D). It's obvious too that if ω were 1 the best protocol would be (C,C), since this is the profile that maximizes U and it would be enacted for sure. But if $\omega < 1$ matters are not so clear; if the protocol is (C,C), then if one agent were to fail, U would go right down to 0. Let us work out the solution. Consider the protocol (C,C). With probability ω^2 both will follow it, (C,C) will be chosen and a value of U of 4 will result; with probability $2\omega(1 - \omega)$ just one will follow it, (D,C) or (C,D) will be chosen and 0 will result; and with probability $(1 - \omega)^2$ neither will and 2 will result. So the expected value of U from (C,C) is $EU(\text{C,C}) = 4\omega^2 + 2(1 - \omega)^2 = 6\omega^2 - 4\omega + 2$. If the protocol is (D,D), (D,D) always gets played, so $EU(\text{D,D}) = 2$. Thus the best protocol, o^{**}, is (C,C) if $6\omega^2 - 4\omega > 0$, or $\omega > 2/3$, and $o^{**} = $ (D,D) if $\omega < 2/3$. More generally, if the parameters of the Prisoner's Dilemma are a, b, s, t instead of 2, 1, −3, 3 the critical value of ω is[12]

$$\omega = \frac{2b - (s + t)}{(a + b) - (s + t)}$$

As we might expect, as agents get more reliable (more likely to function), there comes a point at which the best protocol gets to be (C, C), but below this it is optimal for the mechanism to play safe, ensuring a U value of 2.

One feature of circumspect team reasoning may seem strange. Consider a particular agent P1. Suppose she functions. Since the protocol

o^{**} was formed without the information that P1 is functioning, it might be thought that once P1 knows she is, she should revise her part of it. Consider the Prisoner's Dilemma case: now instead of a probability ω^2 that both will conform, the probability is higher—it's ω. Surely, one may feel, once P1 knows this—and so that it is more likely that the team will be larger—she should for the team's sake consider modifying her behaviour? The trouble with this objection is essentially that P1 has no way of communicating to others that she is in the functioning state and has changed her plan, so they cannot adjust to the change in her plan. P1, though she is doing her best for U, is thinking individualistically, and this is liable to produce a coordination failure. Think in terms of a director: if the director had known that P1 would function rather than attaching probability ω to it, she might well have computed a different protocol; but in it not only P1's part but also others' parts might have been different. P1's independent-minded helpfulness may therefore produce more harm than good. Thinking as a team means among other things keeping to the team plan.[13]

In exact analogy to what went before, we can call a choice mechanism for an unreliable coordination context (S,ω,O,U,f) a *team mechanism* if it brings it about that each i in S chooses o_i^{**} if she functions under the mechanism, that is, a mechanism is a team one if it delivers the protocol o^{**}. As before, a 'team' is a set of agents whose behaviour is governed by a team mechanism; so here, if M is a team mechanism, the *team* is the set of agents who turn out to function under M. Thus a team is a random set of agents.

In an unreliable coordination context, team reasoning is defined as follows. Each agent i in T determines o^{**}, then identifies and executes o_i^{**}. This procedure, team reasoning in an unreliable coordination context, I shall call *circumspect team reasoning*. For circumspect team reasoning to be feasible, all in T must know the payoff structure, the failure probability ω, and the default choices of all. It's clear that circumspect team reasoning is a team mechanism, and that the agents who do it on the day constitute a team. The appropriate version of direction, *circumspect direction*, is now obvious: the director computes o^{**}, defined as just now, and sends each component to the agent in control of it, who executes it.

One example of an unreliable coordination context is a multi-agent system in the sense of computer science, in which the agents are programmed computers known as 'artificial agents', and a failure is a mechanical breakdown. The problem breakdown raises for the design of multi-agent systems is well studied (Fagin et al. 1995). In most applications, either a computer fails, in which case it produces a null output, or it functions according to its design. If the breakdown probability is

$1 - \omega$, the whole situation can be represented by an unreliable coordination context (S,ω,O,U,f) together with the mechanism M, where S is a set of artificial agents, U is the system goal, f is a vector of null outputs, and M is the overall program written for the system.

A simple psychological model of an unreliable coordination context (S,ω,O,U,f) together with the mechanism of circumspect team reasoning goes like this. Label the (circumspect) team-reasoning mechanism TR. The individual human agent i can be in one of two possible psychological states. In one she is somehow moved to reason in accordance with TR; in the other in a standard individualistic way. She functions or fails according to whether she is in the former or the latter state. Although this model is very simple, it is rich enough to express the missing link in the theory of how group identification produces A-choices. What could it be that moves an individual to reason according to TR? In section 4, I will suggest that one TR-priming process—one thing that can move her thus—is group identification. If so, then ω is just the probability that someone group-identifies in the circumstances.[14]

The second limitation of the mechanism of restricted team reasoning is that it fails to address the strategic remainder problem. It assumes that what the failing agents do is 'given', and known to the functioning agents. But the remainder have an agenda of their own, and they may choose strategically to further it, subject to their beliefs about what the team reasoners will do; and in this case there is a noncooperative game between the functioners and the remainder. Consider a situation in which a person may either work for a team (by participating in some team mechanism M designed to maximize some U) or, if not, seek her own ends. For example, suppose $S = \{1,2\}$ are confronted by some two-person game, that if i is in one psychological state she would be motivated to conform to M, for the sake of U, and that if she is in another she would like to maximize her personal utility u_i. Intuitively, there is some sort of three-player game here, for just as the mechanism seeks to optimize U taking into account what agents will do if they fail, so each agent (1,2) who fails will seek to take into account what actions M will produce in the other agent if he does not fail. The difference from an ordinary game is that which of the three players a given human agent is 'serving' is not fixed but determined by some random 'priming' process. The theory of unreliable team interactions in Bacharach (1999) spells out this intuition.

The main lesson we have learned in this section is that there is a mode of reasoning that is efficient for an objective U even when there are agents who will not or may not employ it. It is efficient in the sense that its use by all those available for pursuing U maximizes the expected value of U. And it is in the spirit of simple team reasoning;

indeed, simple team reasoning is just the special case of it which arises when it is universally employed. I have been concerned with efficiency rather than with subjective rationality—with the goodness of the reasons the agents have for deciding in this way. [MB intended to discuss the subjective rationality of team reasoning later in this chapter.]

4. The Reasoning Effect

What produces team reasoning? Team reasoning—in any of its versions—is unmysterious; it is an algorithm, or routine, for arriving at decisions. As many things could get people to execute it as could get them to do long division. It could be done as a game, or as a mathematical exercise, or because someone in control of a group of people found it in her interests that they should.

Among the many scenarios in which people engage in team reasoning, one is of particular interest because it has potentially great explanatory power, in particular the power to explain the facts of Hi-Lo. This is group identification. Although the effects of group identification have been intensively studied, and the studies include effects on the motivation of individuals facing choice problems, there has been no attempt to investigate the *reasoning* that takes place in the heads of group identifiers. But there is an a priori reason to think that, in multiperson decision problems, group identity has the effect of *prompting team reasoning*. I shall call this the *reasoning effect* of group identification.

An agent is an entity that can do alternative things which cause outcomes, which it values more or less. A set of agents defines a set of profiles which cause outcomes; if one entifies the set—thinks of it as a group—and endows it with values, one is thinking of it as an agent. Thus consider a simple coordination context (T,O,U). If one entifies T and endows it with U, one thinks of T as a (complex) agent or—as I shall say for reasons of clarity—as an *agency*. And then one asks questions like 'why did T do that?' or 'what should T do?' Instead of asking 'What should P1 do? What should P2 do?' and so on, as a traditional game theorist asks, she asks '*What should they do?*' in a sense of 'they' that is not equivalent to the first question. Let us see why it is inequivalent. If A is an agent, with a certain option set and a certain ranking of outcomes, the basic normative principle governing 'should-do' claims is that A should do that option in the option set that is highest in the ranking. If one thinks of something as an agent, one is thereby obliged to answer should-do questions about it in this way, by maximizing the goal one attributes to it over the option set one attributes to it. Since in the case of *T* thought of as an agent the option set is *O*, the set of profiles, the should-do question must be to seek the best profile. That is, one must engage in the first step of director reasoning.[15] And this is inequivalent to asking what each member of *T* should do as an individual agent of the kind considered in game theory. To see why it is enough, recall that the former question, about profiles, has a determinate solution, while the latter set of questions does not.

Now suppose that entification is from within. That is, suppose someone self-identifies as a member of the group *T*. Like the outside entifier she might ask herself the question what *T* should do, or why *T* has done something. Like the outsider, she must (in the former case) be asking the question what *T* considered as an agency should do. But because she belongs to *T* the question she must be asking is not this time 'What should they do?' but 'What should *we* do?'

Because such a person is thinking of her *self* as a part of *T*, her conception of the aspect of herself which is her agency undergoes a transformation. We may say that she undergoes not only payoff transformation, now spontaneously wishing to promote *U*, but also 'agency transformation': she thinks of her agential self—her doing and causing self—as a component part of *T*'s agency.

Just as the outside *T*-entifier who thinks about what *T* should do is led to engage in the first step of director reasoning, so the inside entifier who does so is led to engage in the first step of team reasoning. For to ask the question 'What should we do?' is to ask the question what is the best of *T*'s options in terms of *U*; and the options of *T* considered as an entity are the profiles of *O*.

The remainder of the team-reasoning procedure is then inevitable. Once I have computed the best team profile and identified my component in it, team reasoning prescribes that I should choose to perform this component. In the language of chapter 1, section 8.2, team reasoning is a form of profile-based reasoning which has the projection feature. I am rationally obliged to follow the remainder of the procedure. If I believe that *we* should do a certain combination of actions, it is logically required that I also believe that I should do the bit that falls to me. If I am convinced that we should pass each other on the left, I must also think that I should pass you on the left (and that you should do likewise). The underlying general principle is that I cannot coherently will something without willing what I know to be logically entailed by it. This is a standard inference rule of deontic logic, the logic of what ought to be.[16,17]

I have been describing a process in which group identification effects an agency transformation and primes team reasoning. I've been doing so for the simple case in which everyone in the group believes that this mechanism is operating. If we wish, we can easily deal with the case

I discussed in the last section, in which the functioning of the mechanism is unreliable. Variable frame theory would lead us to expect precisely that, in a given situation, group identification happens only with some probability, and those who group-identify are aware of this. The upshot will be that these are led to engage in circumspect team reasoning.

In this world there is some tendency to entify humanity and to see its common aim as the attainment of happiness, and also to understand that not everyone will instantiate this tendency. The theory I have been sketching implies that if and when the entification is from within, that is, the individual sees herself as part of this entified humanity, she will be led through transformations of utility and agency to seek to play her part in the profile of actions that maximizes the expected value of total happiness. This is not cooperative utilitarianism, both because the latter takes the set of followers to be known by each of them and because it is a normative theory, while this is a psychological one. But it describes one process which might plausibly lead people to behave spontaneously according to the precepts of a circumspect version of cooperative utilitarianism.

Group identification is a framing phenomenon. Among the many different dimensions of the frame of a decision-maker is the 'unit of agency' dimension: the framing agent may think of herself as an indivdual doer or as part of some collective doer. The first type of frame is operative in ordinary game-theoretic, individualistic reasoning, and the second in team reasoning. The concept-clusters of these two basic framings centre round 'I/she/he' concepts and 'we' concepts respectively. Players in the two types of frame begin their reasoning with the two basic conceptualizations of the situation, as a 'What shall I do?' problem, and a 'What shall we do?' problem, respectively.[18]

5. Team Reasoning and Collective Intentions

People seem to have 'collective intentions' or 'shared' or 'joint' or 'plural' or 'we' intentions. Gilbert (1989) drew our attention to the fact that 'We intend to pick cherries after breakfast' sometimes seems not to mean just that I intend to and you intend to. Joint intentions present two major problems, to date mainly addressed only in philosophy. First, is 'we intend to' always reducible to I intentions, she intentions, and beliefs? Some writers on the subject (Bratman 1993; Tuomela 1995) think we can construct joint intentions out of these individual-level states; Gilbert (1997, 1998) and Searle (1990, 1995) disagree.

Second, how can people arrive at a joint intention? One way a joint intention apparently arises is that I intend to do x if and only if you do y and you intend to do y if and only if I do x, and these facts are common knowledge. Most discussions focus on situations in which the knowledge of each other's intentions arises through declarations of the conditional intentions, that is, promising to do x if the other does y and so on. However the knowledge arises, this aetiology of joint intentions is mysterious. On the one hand, in the cases in question it seems that the joint intention is 'categorical', not conditional on something; on the other, it seems we can never get rid of the conditions in the individual intentions ('deconditionalize' them). I will if and only if I know you will; but I know that you will if and only if you know I will; hence I will if and only if I know you know I will; hence, similarly, I will if and only if I know you know I know you will; and so on. There is no end to this sequence of implications, and all give conditions for my intention, so this path certainly does not lead to a categorical intention.

I shall argue for two claims. (A) We cannot construct out of individual materials even a weak state of joint intention such as Bratman's; the argument here depends on associating Bratman states with Nash equilibria and the fact that Nash equilibrium states cannot in general be derived in models of individually rational players (see, for example, Bernheim 1984; Pearce 1984; Bacharach 1987). (B) Team reasoning does carry agents to joint intentional states; moreover, these are stronger than Bratman's in that they have features which explain the 'sense of collectivity' that Searle and others think accompany states of joint intention.

It may help readers used to thinking about decisions but not about intentions that much talk about intentions is readily translated into closely matching talk about decisions. A decision is an event that produces an intention, which is a state. Conversely, an intention to do x at time t is normally due to a deliberation that produced a decision at some time before t to do x.

There are several closely related individualistic accounts of what joint intentions are. Here I will use Bratman's (1993) as exemplary. Bratman hopes to capture the vague idea of a joint intention by the following more precise notion, 'shared intention'. Let Z be some outcome that can be produced by P1 and P2, say by the option pair (x, y). Z might be the event that the cherries are picked and x,y the options of picking the low-down ones and picking the high-up ones. Then P1 and P2 have a *shared intention* that Z if: P1's intention is that Z and that Z come about through inter alia P2's intention; P2's intention is that Z and that Z come about through inter alia P1's intention. P2's intention is that Z and that Z come about through inter alia P1's intention. It can also be expressed by an infinite sequence of nested conditions: each intends that Z and that Z will come about through inter alia (the intention of the other that Z and that Z will come about through (the intention of the other that . . .)).

In any two-player game, if the players deliberate to a Nash equilibrium (x,y) of that game, there is a shared intention that Z, where Z is the outcome of (x,y). (I shall explain why in the next paragraph.) Conversely, in any two-player game, if the players deliberate to a state of joint intention, the joint intention must be for a Nash equilibrium outcome. Now it is the case, and increasingly widely recognized to be, that in games in general there's no way players can rationally deliberate to a Nash equilibrium. Rather, classical canons of rationality do not in general support playing in Nash equilibria. So it looks as though shared intentions cannot, in the general run of games, by classical canons, be rationally formed! And that means in the general run of life as well. This is highly paradoxical if you think that rational people can have shared intentions. The paradox is not resolved by the thought that when they do, the context is not a game: any situation in which people have to make the sorts of decisions that issue in shared intentions must be a game, which is, after all, just a situation in which combinations of actions matter to the combining parties.

Nash equilibrium involves shared intention for the following reason. Say (x,y) is a Nash equilibrium and write Z for its relevant consequences. Clearly P1 intends Z, and believes that Z will come about in part because P2 will do y, and that P2 will do y because he intends to, and that P2 intends to do y because he intends Z and because he believes that Z will come about in part because P1 will do x, and so on: we pick up all the clauses in Bratman's condition (assuming that 'intending that Z in part through P2's intention I' implies believing that P2 has the intention I). Or, putting things in terms of the pair of conditions: in this Nash equilibrium clearly P1 intends Z and believes that Z will come about in part through P2's intention; and the counterpart. The converse is easily shown, I think, by backstepping. Say P1 and P2 have a shared intention that Z, through (x,y). Since P1 believes that Z will come about in part through P2's doing y, P1 expects y; since in these circumstances she intends x and she is instrumentally rational, x must be a best reply to y; similarly y must be to x.

Turn to the idea that *a joint intention to do (x,y) is rationally produced in 1 and 2 by common knowledge of two conditional intentions*: P1 has the intention expressed by 'I'll do x if and only if she does y', and P2 the counterpart one. Clearly P1 doesn't have the intention to do x if and only if P2 in fact does y whether or not P1 believes P2 will do y; the right condition must be along the lines of:

(C1) P1 intends to do x if and only if she believes P2 will do y.

(More fully: P1 intends at time t_1 to do x at t_2 if and only if she believes at t_2 that P2 will do y at t_3; or, intends at t_1 to form an unconditional intention to do x at t_2 if and only if she comes to believe at some intermediate time that P2 will do y at t_3.) Label the counterpart condition C2. Call the italicized hypothesis DC. I so label it because it is a way of saying that rational 'deconditionalization' is possible.

So far, I have not assumed that P1 knows C2 or that P2 knows C1. Now suppose P1 comes to know C2 (that P2 intends to do y if and only if he thinks P1 will do x). Suppose that both are instrumentally rational (best-reply rational) and there is common knowledge of this. Then P1 comes to know that y is a best reply to x. Thus, if P1 and P2 come to have common knowledge of C1 and C2, they come to have common knowledge of part of the best-reply mapping of a game. To simplify, suppose it is already common knowledge that each has only two feasible options—P1 must do either x or x', and P2 must do either y or y'. Then they come to know that only x is a best response to y, and that only y is a best response to x, so (x,y) is a Nash equilibrium; and that x' and y' are best responses to each other, so (x',y') is a Nash equilibrium.

On this analysis the effect of learning each other's conditional intentions is to generate a game G with two Nash equilibria. If all the decision-relevant aspects of the situation are features of this game—as I think they are—then DC is correct if and only if, faced with the game G, it is rational to choose (x,y). The 'if' part of this holds because since (x,y) is a Nash equilibrium, if the learning of the conditional intentions rationally leads to (x,y) it leads to a joint intention in the sense of a shared intention.

There are two obstacles to showing this, one superable, the other not, I think. First, there are two Nash equilibria, and nothing in the setup to suggest that some standard refinement (strengthening) of the Nash equilibrium condition will eliminate one. However, I suspect that my description of the situation could be refined without 'changing the subject'. Perhaps the conditional intention C1 should really be 'I'll do x if and only if she'll do y, and that's what I would like best'. For example, if x and y are the two obligations in a contract being discussed, it is natural to suppose that P1 thinks that both signing would be better than neither signing. If we accept this gloss then the payoff structure becomes a Stag Hunt—a Hi-Lo if both are worse off out of equilibrium than in the poor equilibrium (x',y'). To help the cause of rationally deriving the joint intention (x,y), assume the Hi-Lo case. What are the prospects now? As I have shown in chapter 1, there is no chance of deriving (x,y) by the classical canons, and the only (so far proposed) way of doing to is by team reasoning.

This is the outline of my argument that joint intentions are rationally derivable, and in particular are derivable from knowledge of conditional intentions, but that the reasoning that the joint intenders use to get them there must be team reasoning.

The nature of team reasoning, and of the conditions under which it is likely to be primed in individual agents, has a consequence that gives further support to this claim. This is that joint intentions arrived at by the route of team reasoning involve, in the individual agents, a 'sense of collectivity'. The nature of team reasoning has this effect, because the team reasoner asks herself not 'What should I do?' but 'What should we do?' So, to team-reason, you must already be in a frame in which first-person plural concepts are activated. The priming conditions for team reasoning have this effect because, as we shall see later in this chapter, team reasoning, for a shared objective, is likely to arise spontaneously in an individual who is in the psychological state of group-identifying with the set of interdependent actors; and to self-identify as a member of a group essentially involves a sense of collectivity.

6. The Hi-Lo Facts Explained[19]

6.1 *The Effect of Team Reasoning in Hi-Lo*

Sugden noted in 1993 that if players team reasoned in a game of Hi-Lo form, they would choose A. I shall call this property of team reasoning the A-claim. The A-claim suggests a possible resolution of the Hi-Lo paradox. It might be that the empirical tendency to choose A is due to a tendency to team-reason in the circumstances of Hi-Lo. Furthermore, if we are constituted to team-reason, in these circumstances, this would also explain why A seems obviously right to us. This conviction might be all the stronger if team reasoning is a logically valid mode of reasoning.[20] I shall argue that this is the essential truth. But work has to be done both to show that team reasoning is valid, and not fallacious like magical thinking; and to explain why we tend to team-reason when we do, and in particular why we tend to in Hi-Lo.

To team-reason is by definition to team-reason *for* some goal, captured in a team payoff function U. The claim that if a player team-reasons in Hi-Lo she chooses A seems obvious. But it rests on an implicit assumption. This is that the team goal U is simply the shared individual payoff function. So the A-claim should be glossed as: *In Hi-Lo, team reasoning for the shared payoff function gives A.*[21]

Sugden himself proceeds to argue that it is rational to team-reason in Hi-Lo. I shall discuss this normative claim of Sugden's soon. (There is a proviso, relating to your beliefs about how your coplayer is reasoning.

Label this the Sugden proviso.) Sugden's claim would provide, if accepted, support for the judgemental fact about Hi-Lo; it would also provide some for the behavioural fact about Hi-Lo, since rational players might be moved by the same considerations as the rational player Sugden.[22] But I want to suggest a different, though overlapping, account of why people might team reason, and so make A-choices in Hi-Lo. It differs from Sugden's in two ways. One is that, although I shall claim that team reasoning *is* rational, my theory makes the rationality of team reasoning enter less directly than Sugden's in explaining why people team-reason. The immediate explanation of why they team-reason is that they frame their problem in a certain way. The other is that my theory has a rather general scope: the mechanism it describes can work even when the Sugden proviso is not met, and not only in Hi-Lo but also in a much wider range of games of cooperation.

6.2 *A Theory of A-Choice in Hi-Lo*

We have progressed towards a plausible explanation of the behavioural fact about Hi-Lo. It is explicable as an outcome of group identification by the players, because this is likely to produce a way of reasoning, team reasoning, that at once yields A. Team reasoning satisfies the conditions for the mode-P reasoning that we concluded in chapter 1 must be operative if people are ever to reason their way to A. It avoids magical thinking. It takes the profile-selection problem by the scruff of the neck. What explains its onset is an agency transformation in the mind of the player; this agency transformation leads naturally to profile-based reasoning and is a natural consequence of self-identification with the player group.

But the explanation of A is incomplete as yet: the A-claim is no more than an if–then statement, in which group identity is the antecedent—or a mere exogenous assumption. If one day we find reason to expect group identity then, but only then, can we complete the explanation of A-choice via team reasoning. Fortunately, the general theory of group identification allows us to make further progress, and to endogenize the assumption. For the general theory implies, as I argued in chapter 2, that the probability of group identification in a game is a function of identifiable characteristics of that game.

Hi-Lo induces group identification. A bit more fully: the circumstances of Hi-Lo cause each player to tend to group-identify as a member of the group G whose membership is the player-set and whose goal is the shared payoff.

In section 4.1 of chapter 2, I sketched a theory of endogenous group identification in games. I described a feature of interactions I called strong interdependence, defined a hypothesis I called the interdependence hypothesis (IH), which says that strong interdependence is a feature which, if noticed, tends to makes people group-identify, and I argued that IH is implicit in the psychological literature. I argued further that the strength of the tendency to group-identify asserted in IH is determined by the salience of strong interdependence, the absence of salient features of the situation that might militate for individual self-identification, and possibly also by the degree of the strong interdependence. Finally, I argued that if and when endogenous group identification takes place, the group utility function will be Paretian and, a fortiori, satisfy Unanimity.

In Hi-Lo the strong interdependence of the players is highly salient, and there are no obvious countervailing features; in particular, there is perfect harmony of interest between the players. Since in Hi-Lo the players' payoffs coincide, the group payoff function is, by Unanimity, just the common individual payoff function. All this implies the following claim, in which P is the player set and U the common individual payoff function. Claim I is intended only as a rough quantitative claim, and I make no attempt to remove the vagueness of the word 'strong' in it.

(Claim I) In Hi-Los the tendency for a player to identify with (P,U) is strong.

Claim I, together with the reasoning effect (group identity stimulates team reasoning) and the A-claim (in Hi-Lo team reasoning for the shared payoff yields A), explains the behavioural fact that people tend to play A.

If what induces A-choices is a piece of reasoning which is part of our mental constitution, we are likely to have the impression that choosing A is obviously right. Moreover, if the piece of reasoning does not involve a belief that the coplayer is bounded, we will feel that choosing A is obviously right against a player as intelligent as ourselves; that is, our intuitions will be an instance of the judgemental fact. I suspect, too, that if the reasoning schema we use is valid, rather than involving fallacy, our intuitions of reality are likely to be more robust. Later I shall argue that team reasoning is indeed nonfallacious. [MB intended to present this argument later in this chapter. The editors attempt to reconstruct the argument in section 1 of the conclusion.] And so the judgemental fact is explained too, as long as Claim I is not only true but also perceived, by those affected by it, to operate on others. But this last condition is also likely to fulfilled, for it is merely an application of a general fact about framing [that MB would have explained in chapter II].

When framing tendencies are culture-wide, people in whom a certain frame is operative are aware that it may be operative in others; and if its availability is high, those in it think that it is likely to be operative in others. Here the framing tendency is—so goes my claim—universal, and a fortiori it is culture-wide.

To summarize where we have got to: in a Hi-Lo, a human is likely to think about herself and her coplayer as 'us', to ask 'what should *we* do?' and to expect her coplayer to be asking himself the same question. Since it is a theorem that what we should do is (A,A), and hence that if I am part of us I should do A, reaching the conclusion that I should do A seems obviously rational to me, and its rationality to me does not depend on my imputing irrationality to my coplayer.

And so we have at last the outline of an explanation of the facts of Hi-Lo. The heart of this explanation is endogenous group identification in the player set, and the stimulation by group identity of team reasoning.

6.3 Extending the Endogenous Group Identity Theory

The explanatory potential of team reasoning is not confined to pure coordination games like Hi-Lo. Team reasoning is assuredly important for its role in explaining the mystery facts about Hi-Lo; but I think we have stumbled on something bigger than a new theory of behaviour in pure coordination games. The key to endogenous group identification is not identity of interest but common interest giving rise to strong interdependence. There is common interest in Stag Hunts, Battles of the Sexes, bargaining games and even Prisoner's Dilemmas. Indeed, in any interaction modelable as a 'mixed motive' game there is an element of common interest. Moreover, in most of the landmark cases, including the Prisoner's Dilemma, the common interest is of the kind that creates strong interdependence, and so on the account of chapter 2 creates pressure for group identification. And given group identification, we should expect team reasoning.

But for the theory of endogenous team reasoning there are two differences between the Hi-Lo case and these other cases of strong interdependence. First, outside Hi-Los there are counterpressures towards individual self-identification and so I-framing of the problem. In my model this comes out as a reduction in the salience of the strong interdependence, or an increase in that of other features. One would expect these pressures to be very strong in games like Prisoner's Dilemma, and the fact that C rates are in the 40 per cent range rather than the 90 per cent range, so far from surprising, is a prediction of the present theory. I will come back to this when I address the problem of cooperation as it is traditionally posed, as the questions of whether Pareto-optimality can be rationally achieved, and how it is ever achieved, in the Prisoner's Dilemma and other adverse-incentive games. [This would have been the subject of chapter IX. In section 3 of the conclusion, the editors try to reconstruct what MB intended to say.]

The second difference is that outside coordination games we need a theory of how U (the perceived payoff of the identified-with player set) relates to the individual payoffs. As I argued in chapter 2, everything we know about group identification suggests that any payoff U that tends to prime identity with the player set for U will be Paretian in u_1 and u_2. Beyond this we don't know much yet, but it is reasonable to suppose that principles of symmetry between individual payoffs will be respected in U.

7. Evidence for Team Reasoning

I have suggested a radical hypothesis about how people reason in interactive situations: that people sometimes spontaneously team-reason.[23] In the last section I presented a theory of play in Hi-Lo and other games in which people are led by group identification to team-reason spontaneously. But what is the evidence for thinking that, even when people group-identify, they spontaneously team-reason? There are five kinds of evidence, all of which support this hypothesis—the hypothesis that there is a Reasoning Effect of group identification. Since group identification itself is a well-established phenomenon, this evidence is evidence a fortiori that humans sometimes spontaneously team-reason.

First, there is the logical connection between group identity and team reasoning we saw earlier. If a group identifier thinks of herself as part of a 'we'—self-identifies with 'us'—it is only for 'us' that she can intelligibly deliberate.

Second, there is introspective evidence. In the Unlocking example of chapter 1, I was driving one of the cars. My own strong impression immediately afterwards was that I had reached my own decision what to do just when I realized that the only way we could deal with the jam efficiently was to move in the sequence 2, 4, 3, 5, 1. I also had the distinct impression that this realization had been produced by a search for a good way to deal with the jam—for us to deal with it. Examples of this kind of introspective impression are easy to multiply.[24]

Third, there is evolutionary evidence. There is reason to believe that team reasoning is a component part of the proximate mechanism in man for efficient group decision-making. In Chapter 3 I argued that we are likely to have evolved to be good at group decision-making, and that the evolved proximate mechanism for this decision-making skill is likely to involve group identification. I claimed in support that group identity, unlike altruism, can produce cooperation not only in adverse-incentive but also in common-interest games; and I promised to say in the present chapter how it could do this. We have now seen that it would if group identity prompts team reasoning, that is, if human psychology includes a Reasoning Effect. Thus if we have reason to believe in a Reasoning Effect (such as the logical reason rehearsed above), evolutionary considerations back group identity (and so team reasoning).

There is a second evolutionary argument in favour of the spontaneous team-reasoning hypothesis. Suppose there are two alternative mental mechanisms that, given common interest, would lead humans to act to further that interest. Other things being equal, the cognitively cheapest reliable mechanism will be favoured by selection. As Sober and Wilson (1998) put it, mechanisms will be selected that score well on availability, reliability and energy efficiency. Team reasoning meets these criteria; more exactly, it does better on them than the alternative heuristics suggested in the game theory and psychology literature for the efficient solution of common-interest games. The classical solution is to begin by computing all the Nash equilibria, then select among them, and in general Nash equilibrium computation is an order of magnitude more complex than determining the best profile.[25] Even bounded rationality solutions, such as the Stackelberg heuristic, are computationally much more complex.

Fourth, there is transcendental evidence: the argument that there must be team reasoning because only it can explain something. The team-reasoning theory reaches parts that no other theory can reach. This argument has two forms, one modest and one grandiose. The modest argument is that all existing theories fail, while the team-reasoning theory succeeds. This has been the burden of chapter 1 and the present chapter. Existing theories that purport to explain the behavioural fact, the fact of A-play, are either question-begging or problem-transforming or appeal to irrationalities and so do badly at explaining the judgemental fact. The grandiose argument is that there could not possibly be a theory yielding A that did not imply team reasoning. I argued in this direction at the end of chapter 1 when I argued that any theory that both yields A and responds to what seem to be our intuitions about why A is justified must involve mode-P reasoning.

Finally, there is experimental evidence. There are well-known and powerful objections to relying on self-reports in determining how decision-makers are reasoning.[26] We need a behavioural test. But devising a behavioural test presents unusual difficulties in the present case.

The main obstacle to a behavioural test is that if inducing group identification increases observed cooperation, this can be explained by payoff transformation without any change in the way people reason. For example, identification by Stag Hunt players turns the payoff structure into a Hi-Lo, so any theory of A-play whatsoever suffices to explain an increased S rate. A test using Hi-Lo games is proof against this difficulty, for we can be sure there is no payoff transformation since payoffs are already identical. But the problem is that a group-identification manipulation would be hard put to make any significant difference to the A rate, as observed A rates are usually near 100 per cent anyway.

Bacharach and Guerra [in unpublished experimental work] found a way around these difficulties, and devised a behavioural test of the hypothesis that group identification raises cooperation through prompting team reasoning, against the leading alternative, that it does so by transforming payoffs while leaving the form of reasoning unaltered. As we know, if group identifiers in Hi-Lo are standard, best-reply reasoners, then they face an equilibrium selection problem. Harsanyi and Selten (1988) argue that they are then led to play A because its optimality makes it salient. We turn this idea on its head. If what makes standard reasoners play A is just that it is *salient*, then it is possible in principle to get them to play B, by making B even more salient! On the other hand, team reasoners face no equilibrium selection problem, and so their behaviour should be unaffected by raising the salience of B. The Reasoning Effect would be established if we could first present a Hi-Lo in a way which makes B salient enough to produce a significant amount of B-play, and if we then observed that after a group-identity manipulation this same presentation produced a significantly lower amount of B-play.

[Subjects played twenty two-player games with a succession of unknown partners. Four of these, called tasks 10, 12, 17 and 18, were classic Hi-Lo matching games. In task 10, each subject was asked to choose one of four items in a display; the prize for choosing the same item as the partner was £6, £5, £4 or £3, depending on the item chosen. The display is shown in figure 4.3. In this task, subjects won the highest prize by matching on the ace of hearts, but the £5 item (the ace of spades) was particularly salient. Task 12 had the same form, but the items were flower names. The £5 item, 'rose', was made salient by reporting the 'social history' of choices in essentially the same matching experiment (Mehta, Starmer and Sugden 1994), when 'rose' was easily the most common choice. Tasks 17 and 18 were Hi-Los with only two items, one with a prize of £6 and one with a prize of £5; the £5 item was made salient. (In these tasks there were nonzero but smaller prizes for nonmatching choices.)

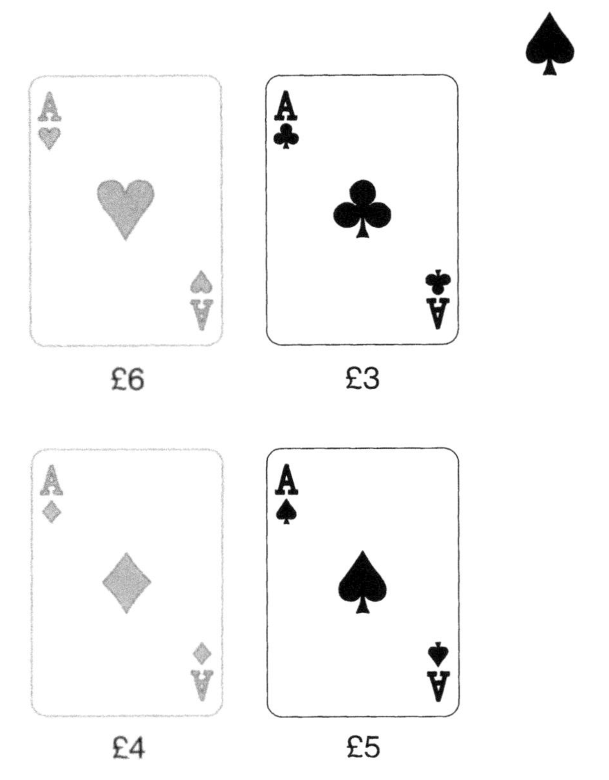

Figure 4.3. Task 10 display

These tasks were presented in slightly different forms to two different groups of thirty-two subjects. In one group ('group G') but not the other ('group N') subjects were subjected to a number of group identity manipulations. In describing tasks to group G, we used more group-oriented language (for example, referring to the other player as 'your partner' rather than 'the other person'). We told group G subjects (depending on the composition of their session of the experiment) that they were all undergraduates, or all people who lived and worked in Oxford, or whatever. We imposed a common fate: for group G subjects, the payments shown in the displays were multiplied by three if a tossed coin landed heads.[27]

The results are shown in table 4.1. In all tasks, in both groups, the majority of subjects chose the £6 item. However, a sizeable minority chose the salient £5 item, while the other items were almost never chosen. This evidence supports the conjecture of Harsanyi and Selten and

TABLE 4.1.
Results of experiment

	Group N (without group-identity manipulations) number of subjects choosing			Group G (with group-identity manipulations) number of subjects choosing		
Task	£6 item	£5 item	Other	£6 item	£5 item	Other
10	22	10	0	23	9	0
12	23	8	1	25	6	1
17	19	13	n.a.	24	8	n.a.
18	18	14	n.a.	27	5	n.a.

N.a.: Not applicable.

others that, when equilibrium is not unique, players who reason in a standard game-theoretic way are swayed by salience, even if the salient equilibrium is Pareto-inferior to some other equilbrium.[28]

The most interesting comparisons are between the two groups, isolating the effect of the group-identity manipulations. In each task, the proportion of subjects choosing the £6 item—that is, choosing their components of the strategy profile that is best for the two players as a team—is higher in the presence of the group-identity manipulations (group G) than in their absence (group N); the average value of this proportion is 70 per cent in group G and 64 per cent in group N.[29] This is a preliminary study, on a modest scale in terms of the number of participants, and subject to the caveats that apply to all statistical hypothesis tests, and to all laboratory experiments. But it provides some behavioural evidence for the hypothesis that group identification induces spontaneous team reasoning.]

8. Variable Identity

The reasons why group identifiers team-reason do not depend on what causes the identification.[30] It does not have to be endogenous, caused by features of the decision situation itself. In particular, it does not have to be generated by the agents' individual preferences. When two English explorers meet by chance in the African wilderness, they stand ready to die for each other not because of strong interdependence but because they went to the same school. The group identities that people have when they make decisions, and the group payoffs that go with them, can be, and in fact typically are, quite exogenous to the decision problem that they face.[31]

The psychology of group identity allows us to understand that group identification can be due to factors that have nothing to do with the individual preferences. Strong interdependence and other forms of common individual interest are one sort of favouring condition, but there are many others, such as comembership of some existing social group, sharing a birthday, and the artificial categories of the minimal group paradigm.

But recall that *we* are interested in interactive decisions. It might be thought that here all does depend on individual preferences over possible outcomes. But this not so. Several of these alternative sources of group identity have been used by experimentalists to try to induce group identity in subjects facing interactive decisions; and many of these attempts were successful. Moreover, in these experiments the individual preferences were not concordant—typically the tasks were Dilemmas—so it cannot have been that the experimentalists' manipulations were doing nothing and the group identity was in fact produced by the individual payoff structure. Formally, what is going on in these experiments is that the mixed-motive game defined by the individualistic payoffs is getting replaced, owing to the group-identity manipulation, by one in which the players have a group identity and the payoffs are that of the group.

Wherever we may expect group identity we may also expect team reasoning. The effect of team reasoning on behaviour is different from that of individualistic reasoning. We have already seen this for Hi-Lo. [MB intended, later in the book, to show that this is also true of other forms of interaction. The implications of team reasoning for the Prisoner's Dilemma are discussed in section 3 of the conclusion.] This has wide implications. It makes the theory of team reasoning a much more powerful explanatory and predictive theory than it would be if it came on line only in games with the right kind of common interest. To take just one example, if management brings it about that the firm's employees identify with the firm, we may expect them to team-reason and so to make choices that are not predicted by the standard theories of rational choice.

In games in which the experimenter tries to induce group identity, and in naturally arising scenarios in which people group-identify, identification is almost never certain. At most there is *some pressure* for self reidentification. This pressure is almost never overwhelming, but it raises to some degree the probability of a reidentification. Recall that different identities are different frames. Whether the factors favouring group identification are experimental manipulations or occur naturally in the real social world, like factors favouring different framings in other zones of framing, they can only produce tendencies. As with the framings of physical objects in the variable frame theory of focal points, I shall therefore model them as giving rise to probability distributions over alternative frames, in this case identity frames. [MB was planning to come back to this in a later, unwritten section.]

As we have seen, the same person passes through many group identities in the flux of life, and even on a single occasion more than one of these identities may be stimulated. So we will need a model of identity in which the probability of a person's identification is distributed over not just two alternatives—personal self-identity or identity with a fixed group—but, in principle, arbitrarily many.

Notes

1. The interdependence comes from this feature of the setup. How much I am *paid* for a unit of my effort depends on others' effort. There need not be interpendence in the production function: the task could be 'additive' rather than having 'conjunctive', 'disjunctive' or some other form of technical interdependency.

2. Sugden calls this 'thinking as a team' in (1993) and 'team-directed reasoning' in (2000). I called it 'team reasoning' in (1999) and (2001).

3. I am assuming, and shall continue to assume for some time, that U has only one maximizer. Most situations in the field are like this. The issues that arise in the case of multiple maxima are not trivial, but going into them would be a distraction at this point. Suffice it to say that the same considerations about the relative salience of different descriptions that reduce coordination games with several maxima to ones with unique maxima, which we met in chapter 1, have application here.

4. The first computation is just maximizing a function of an n-tuple. Suppose $n = 1{,}000$ and $i = 631$, so i knows she controls the 631st element. She still has to count up to 631, or backwards 369 places, or use some other algorithm to locate the thing she has to do.

5*. MB intended to explain that it is important for the subjective rationality of team reasoning that the payoff structure of a game is not merely known to all players but is common knowledge. However, he planned to leave discussion of this until a later section of this chapter, which had not been written at the time of his death.

6. Except that I continue to assume that U has a unique maximizer.

7*. MB intended to refer the reader to the theory of the second best in welfare economics. On this see Lipsey and Lancaster (1956).

8. At least for many possible profiles o in O.

9. We are reluctant to say that Owen is not in the team. It would be different if his motivation were not a team one, if he were feeling lazy or seeking glory. We would perhaps say 'the effective team is down to ten'. My 'teams' are 'effective teams'. I think the reason for the reluctance is that we think of teams in

contexts like football as enduring things, and we expect Owen to be around and playing an active part most of the time over the long run. Such considerations aren't really relevant in the present theory, which abstracts from intertemporal considerations.

10. In game theory it is held that what it is rational for i to do is to act so as to promote her utility u_i; but typically what act this is depends on what others do, so it is unclear what act of yours will best promote u_i; it is an empirical question whether others are rational, and, if so, whether they resolve the ambiguity in the same way. If everyone has the same aim U, whether this is total utility or satisfying the whims of the king, the problem of defining the rational is exactly analogous to that of defining the right.

11. It also leads to *utilitarian generalization*. Rule utilitarianism says that agents should follow a set of rules which it would be best for all to follow. Utilitarian generalization substitutes: each should do whatever would be best for him and all *similarly situated* to do.

12. Note that the problem is interesting only if $2b > s + t$. If the reverse holds, then for all values of ω, even $\omega = 0$, the best protocol is (C,C), because C is dominant for the designer of a choice mechanism for U.

13. Someone might concede that changing on finding that you are functioning might sometimes do more harm than good, but still wonder whether it could help in some cases. Not even this is so. Suppose it were, that is, that some i could raise the conditional expectation of U given the event that he is functioning by changing from o^{**} to, say, o_i'. Then the director could have anticipated this and should have instructed o_i', not o_i^{**}. The director already factors into the protocol the possibility that i will function: she works out the best thing for i to do if he finds he is functioning. All the work has been done; there is no more for i to do than to play her part in the protocol. Another way of seeing that revision is not on is this. Changing to o_i' can only raise the expected value of U on some hypothesis by i as to other agents' behaviour. One such hypothesis is that for each $j \neq i$ the choice of j is unchanged if she functions. But if i's reason for changing is sound, then so is that of all j who function and reason in this way. And if all reason in this way they are reasoning individualistically (for the sake of U); so i must ask herself what conclusions this will lead them to, they will have to ask themselves like questions, and we have got back to the hopeless coordination problem of game theory.

14. This assumes that group identification, if it happens, primes TR with probability 1. More generally, if this probability is $p < 1$, and that of group identification is v, then $\omega = pv$.

15. There may be ways of discovering the answer to the question what is the best profile for T which proceed element-wise, for example asking what is best for element 1 given assumptions about elements 2 . . . . n, then asking about element 2 and revising the answer for element 1, and so forth. But it remains true that in order to answer the question what T should do one must discover the best profile for T. Saying that one team-reasons does not say anything about her computational strategy for identifying o^*.

16. From the premise $A \rightarrow B$ one may infer that $OA \rightarrow OB$, where O stands for 'it ought to be the case that'. The logic of rational belief, 'epistemic logic', tells

us that if a rational believer believes the premise, and that *OA*, she believes that *OB*.

17*. MB was planning a paragraph here on why footballers team-reason.

18. The concept-clusters of alternative frames do not have to be disjoint. Here, no 'we' thought, but only 'I/he/she' thoughts, figure in the standard game-theoretic frame, but some 'I' thoughts figure in the team-reasoning frame, since after deciding what 'we' should do in the computing phase, the decision-maker engages in 'I' thoughts in addressing and instructing herself, that is, in inferring from what the team should do to what she should do.

19*. MB planned a section before this one, entitled 'Properties of Team Reasoning', but his notes for it were too rough to include here. The following sections have been renumbered accordingly.

20. There is some evidence that intuitions supported by valid reasoning are more firmly held than fallacious intuitions. [MB planned to elaborate on this.]

21. It can easily be seen that a stronger claim is also true: in Hi-Lo team reasoning for any *U* that is Paretian gives A. And what is true of Hi-Lo is true of all common-interest games: in any common-interest game, team reasoning by all players for any Paretian *U* yields the unique Pareto-optimal outcome. For example, in Stag Hunt any Paretian group payoff leads team reasoners to choose S. This theorem is a trivial consequence of the definitions of the various terms. A common-interest game has a unique Pareto-optimal profile. Since *U* is Paretian, its maximizer o^* is this profile; so each team reasoner plays her part in o^*. In games that are *not* common-interest, such as Dilemmas, it is still true that team reasoning by all for any given Paretian payoff function yields a Pareto-optimal outcome. The difference between common-interest games and other games is that in the former any Paretian payoff gives the same point (the unique Pareto-optimal one), and in the latter different ones give different points on the Pareto frontier, some of which may be extremely unequal.

22*. Sugden does not claim that rationality *prescribes* team reasoning in Hi-Lo. He argues that rationality must be defined with respect to a unit of agency, and that the individual and the group are equally legitimate units. *If* it is common knowledge that all members of the relevant group conceive of rationality in terms of group agency (the condition that MB calls the 'Sugden proviso'), *then* it is rational for each member to act according to the prescriptions of team reasoning. Symmetrically, if it is common knowledge that all individuals conceive of rationality in terms of individual agency, then it is rational for each individual to act according to the prescriptions of standard game-theoretic reasoning.

23*. MB intended this section to include a discussion of an experiment carried out by Gerardo Guerra and himself, but this part of the manuscript was incomplete. The sections in square brackets are based on the manuscript, but have been expanded by the editors using material supplied by Guerra.

24*. MB planned to cite more reports of this kind.

25*. MB planned a note on the computational complexity of Nash equilibrium.

26*. MB planned a note listing problems with self-reports about reasoning.

27. Subjects were told that, in the case of a tail, the displayed payment would go to the pair, between whom it would be shared equally; in the case of

a head, three times this amount would be paid in this way. Thus if £3 appeared below an object and both chose this object, the pair's expected payment would be £6. The basic payments in the displays were half those in condition 1. Thus we equalized the (expected) payment for coordinating to a *pair* in condition 2 with the payment for coordinating to an individual in condition 1.

28. The evidence supports this view, rather than establishing it. Players not group-identifying, but not reasoning in a standard game-theoretic way either, might also be swayed by salience; for example, 'level-*n*' players might be drawn to the most salient choice in the way described by Bacharach and Stahl (2000). Nevertheless, the Harsanyi-Selten claim is supported, since we have growing evidence (for instance Stahl and Wilson 1994) that a fair proportion of subjects in games are 'Nash players'.

29. This difference is statistically significant (at the 1 per cent level) for one of the four tasks, namely task 18.

30*. As part of this section MB was planning to discuss 'conflicting loyalties' between multiple, shifting group identities.

31. Sugden (1999) contemplates a theory in which agents may have group identities—and use team-directed reasoning—and the group payoff functions are exogenous assumptions of the theory. Their exogeneity might be taken as an objection to such a theory—isn't it 'ad hoc' and so theory-weakening to invoke such things? Sugden points out that the exogeneity of the group preferences is no more unacceptable than the exogeneity of individuals preferences in standard choice theory. However, it must be conceded that in a theory in which people are ascribed both individual and group preferences, unless the latter are endogenized, for example by deriving them from the former, there are more primitives, and this may seem a pity if one likes theories with few primitives. My own response is, first, that the world is a complex place, and that the theorist must take the world as she finds it. Second, that what matters for good theorizing is not the number of parameters or other primitives, but parsimony, which falls with more primitives but rises with more explanatory power. If by adding to n individual preference relations an $(n + 1)$st one can explain cooperation, that is a small price for a great gain, and parsimony increases.

Conclusion

WE, THE EDITORS, must now pick up the threads of the argument. In the form in which it now appears, chapter 4 ends abruptly. Bacharach was working on this chapter—chapter VII in his plan—when he died, and its final sections were never written. It is not clear from his notes exactly what additional material he intended to include in the chapter, but we do know that he was planning a section on 'properties of team reasoning', which would have provided an overview or stocktaking of his ideas on that subject. As we explained in the preface, chapter VII would have been followed by a chapter on communication and a chapter on the Prisoner's Dilemma. He had also considered writing a further chapter on 'the person as a team'. In this conclusion, we try to reconstruct the main ideas that would have been presented in the missing sections of chapter VII and in the subsequent chapters.

In section 1, we present Bacharach's ideas about team reasoning in the form of schemata for practical reasoning, using the analogy of syllogisms in philosophical logic. In his notes for the unwritten chapter VIII, he sets out one such schema explicitly (closely related to the schema that we will call 'basic team reasoning'). We do not know whether he intended to present corresponding schemata for the various forms of team reasoning discussed in chapter 4. Nor can we be sure that, had he presented such schemata, they would have had the same forms as our reconstructions. However, it is fundamental to Bacharach's approach that, for the purposes of game-theoretic analysis, rationality is to be understood as valid reasoning. Given this, it seems that the basic structure of his theory is most transparently expressed in terms of schemata of practical reasoning. By expressing his ideas in this way, we are able (in section 2) to compare his analysis of team reasoning with related analyses proposed by other writers. In section 3, we reconstruct Bacharach's analysis of the Prisoner's Dilemma, completing the project that was the starting point for the book—solving the 'three puzzles of game theory'. This analysis is a fairly straightforward application of the ideas developed in chapters 1 to 4; we are reasonably confident that our treatment of this topic is faithful to Bacharach's intentions. In section 4, we locate this analysis in the literature of game theory.

The remainder of the conclusion is more speculative. It deals with topics on which, as far as we can tell, Bacharach's ideas were still in flux at the time of his death. Section 5 discusses his ideas on the role of communication in team reasoning. Section 6 deals with team reasoning in games in which the players act sequentially—a case which is crucial for Bacharach's analysis of personal identity. It seems that Bacharach was developing an argument which was intended to challenge the validity of backward-induction reasoning—one of the core principles of the conventional analysis of sequential-move games and of rational individual choice over time. We try to reconstruct this argument from scattered clues in his notes. Finally, in sections 7 and 8, we discuss Bacharach's ideas about personal identity.

1. Formalising Team Reasoning

In the introduction, we presented the following schema of practical reasoning:

Schema 1: Individual rationality
(1) I must choose either *left* or *right*.
(2) If I choose *left*, the outcome will be O_1.
(3) If I choose *right*, the outcome will be O_2.
(4) I want to achieve O_1 more than I want to achieve O_2.

I should choose *left*.

According to Bacharach, this reasoning is *valid* by virtue of its being 'success-promoting'. That is, if the premises (1) to (4) hold, then the action commended by the schema's conclusion promotes the agent's objectives. This is true, whatever actions are represented by *left* and *right*, and whatever outcomes are represented by O_1 and O_2.

But now consider Schema 2, in which (*left, right*) denotes the pair of actions 'I choose *left*, you choose *right*':

Schema 2: Collective rationality
(1) We must choose one of (*left, left*), (*left, right*), (*right, left*) or (*right, right*).
(2) If we choose (*left, left*) the outcome will be O_1.
(3) If we choose (*left, right*) the outcome will be O_2.
(4) If we choose (*right, left*) the outcome will be O_3.
(5) If we choose (*right, right*) the outcome will be O_4.
(6) We want to achieve O_1more than we to achieve O_2, O_3 or O_4.

We should choose (*left, left*).

Is *this* valid? Given the symmetries between the two schemata, it seems that, if either is valid, so too is the other. Of course, a decision theorist might reply that 'I' and 'we' are asymmetrical by virtue of the legitimacy of 'I' as a unit of agency and the illegitimacy of 'we'. But such a response begs the question: what makes 'I' legitimate and not 'we'? Bacharach rejects the idea that, merely by fiat, propositions about collective agents can be disallowed. It is central to his analysis of team reasoning that Schema 2 is valid in just the same sense that Schema 1 is.

Yet at first sight, the two schemata seem to be potentially contradictory. For example, consider the Prisoner's Dilemma. Using a variant of Schema 1, each player can reason to 'I should choose *defect*'. But in terms of 'our' objectives, it seems clear that we want the outcome resulting from (*defect, defect*) less than we want that resulting from (*cooperate, cooperate*). Given that premise, the inference to the conclusion 'We should not choose (*defect, defect*)' seems to involve no more than a very natural extension of Schema 2. Bacharach resolves this apparent contradiction by arguing that, properly understood, the two sets of premises are mutually inconsistent. The premises of Schema 1 presuppose that *I* am an agent, pursuing *my* objectives. Those of Schema 2 presuppose that *we* make up a single unit of agency, pursuing *our* objectives. I cannot simultaneously think of myself both as a unit of agency in my own right and as part of a unit of agency which includes you. More generally: instrumental practical reasoning presupposes a unit of agency.

We can make this feature of practical reasoning more transparent by writing schemata in forms which include premises about agency. Consider any situation[1] in which each of a set S of individuals has a set of alternative *actions*, from which he must choose one. A *profile* of actions assigns to each member of S one element of his set of alternative actions. For each profile, there is an *outcome*, understood simply as the state of affairs that comes about (for everyone) if those actions are chosen. We define a *payoff function* as a function which assigns a numerical value to every outcome. A payoff function is to be interpreted as representing what some specific agent wants to achieve: if one outcome has a higher numerical value than another, then the relevant agent wants to achieve the first more than he (or she, or it) wants to achieve the second. Now consider any individual *i*, and any set of individuals *G*, such that i is a member of G and *G* is a weak subset of *S*. We will say that *i identifies with G* if *i* conceives of *G* as a unit of agency, acting as a single entity in pursuit of some single objective. Finally, we define *common knowledge* in the usual way: a proposition *x* is common knowledge in a set of individuals *G* if: (i) *x* is true; (ii) for all individuals *i* in G, *i* knows *x*; (iii) for all individuals *i* and *j* in *G*, *i* knows that *j* knows *x*; (iv) for all individuals *i*, *j*, and *k* in *G*, *i* knows that *j* knows that *k* knows that *x*; and so on.

Letting A stand for any profile and U for any payoff function, consider the following schema:

Schema 3: Team reasoning (from a group viewpoint)
(1) We are the members of S.
(2) Each of us identifies with S.
(3) Each of us wants the value of U to be maximised.
(4) A uniquely maximises U.

Each of us should choose her component of A.

This schema seems to capture the essential features of Bacharach's conception of team reasoning. Notice that, because of (2), the schema does not yield any conclusions unless all the members of S identify with this group. Notice also that we can apply Schema 2 in cases in which S contains only one individual. In this case, S can be written as {myself}. Premise (1) then becomes 'I am the only member of the set {myself}'. Premise (2) reduces to 'I identify with {myself}', which amounts to saying that the reasoning individual views herself as an agent. And then the schema represents straightforward practical reasoning by an individual agent. Thus Schema 3 encompasses both individual and team reasoning.

Schema 3 represents a mode of reasoning that can be used by people *as a group*. What does it mean for a number of people to reason as a group? One way to make sense of this is to imagine those people in an open meeting, at which each of a set of premises is announced, and acknowledged as true, by each person. Then the inference to be drawn from those premises is announced, and acknowledged as valid, by each person. In such a setting, it is common knowledge among the members of the group that each of them accepts the relevant premises. That this is common knowledge does not need to be stated explicitly in the schema; it is not an additional premise, but a presupposition of the whole idea of 'reasoning as a group'.

However, as Bacharach makes clear in chapter 4, team reasoning may also be done by an individual team member. In this case, it is important that the team reasoner decides to do her part in the team action *because* it is her part in the best possible combination for the team. In a schema of practical reasoning, the premises can be interpreted as the agent's reasons for choice; the conclusion says what the agent should do *because of* those reasons. When an individual engages in team reasoning, the premises from which she reasons must make reference to the profile of actions that best achieves the team's goal. If we adopt this approach, Schema 3 can be rewritten as follows:

Schema 4: Team reasoning (from an individual viewpoint)
(1) I am a member of *S*.
(2) It is common knowledge in *S* that each member of *S* identifies with *S*.
(3) It is common knowledge in *S* that each member of *S* wants the value of *U* to be maximised.
(4) It is common knowledge in *S* that *A* uniquely maximises *U*.

I should choose my component of *A*.

In moving from Schema 3 to Schema 4, the conclusion has changed from 'Each of us should choose her component of *A*' to 'I should choose my component of *A*'. Where Schema 3 represents *reasoning by* 'us', Schema 4 represents *reasoning by* 'me'. In both cases, however, the *reasons* refer to 'us'. In section 2 of chapter 4, Bacharach says that there are two levels involved in team reasoning. The first is reasoning at a group level, identifying the profile of actions that is best for the group. The second is to reason as an individual that, because her component of the output of the first stage of reasoning is her component of the best profile, she ought to perform it. Bacharach insists that there is no way to eliminate the reference to 'we' in the reasoning, or to analyse that 'we' in terms of individual intentions. But the locus of action is the individual and the joint intention to do something as a team issues in an action taken by each individual agent (see chapter 4, section 5).

The role of common knowledge in Schema 4 requires some explanation, given that Bacharach almost never refers to common knowledge in chapter 4, except when discussing the work of other theorists. In classical game theory, it is a standard background assumption that the properties of the game under analysis are common knowledge among the players. If players are uncertain about any feature of a (putative) game, that uncertainty is modelled by expanding the game to include 'moves of nature'—that is, chance events. The expanded game is then assumed to be common knowledge.[2] We take it that, following the conventions of classical game theory, Bacharach expects his readers to assume, unless the contrary is stated explicitly, that the properties of the games he analyses are common knowledge among the relevant players. However, when game-theoretic rationality is represented in schemata of practical reasoning to be used by individuals, common knowledge assumptions need to be stated explicitly. And this is exactly what Bacharach does when (in his notes for chapter VIII) he formulates the schema that we represent below as Schema 7.[3]

Recall that Bacharach analyses two variant forms of team reasoning which apply in cases in which it is not common knowledge that all members of S can be relied on to identify with the group. The first of these is restricted team reasoning, which applies in cases in which it is known that certain specific members of S do not so identify. This can be represented by a generalisation of Schema 4. Let T be any subset of S, and let A_T be any profile of actions for the members of T. Then:

Schema 5: Restricted team reasoning
(1) I am a member of T.
(2) It is common knowledge in T that each member of T identifies with S.
(3) It is common knowledge in T that each member of T wants the value of U to be maximised.
(4) It is common knowledge in T that A_T uniquely maximises U, given the actions of nonmembers of T.

I should choose my component of A_T.

The other variant form of team reasoning is circumspect team reasoning, which applies in cases in which, for each member of S, there is some nonzero probability that she fails to identify with the group. This can be represented by a different generalisation of Schema 4. Suppose there is a random process which, independently for each member of S, determines whether or not that individual is a member of the subset T; for each individual, the probability that she is a member of T is ω, where $\omega > 0$. We define a proposition x to be *T-conditional common knowledge* if: (i) x is true; (ii) for all individuals i in S, if i is a member of T, then i knows x; (ii) for all individuals i and j in G, if i is a member of T, then i knows that if j is a member of G, then j knows x; and so on. (As an illustration: imagine an underground political organisation which uses a cell structure, so that each member knows the identities of only a few of her fellow members. New members are inducted by taking an oath, which they are told is common to the whole organisation. Then, if T is the set of members, the content of the oath is T-conditional common knowledge.) We define a *protocol* as a profile of actions, one for each member of S, with the interpretation that the protocol is to be followed by those individuals who turn out to be members of T. Let P be any protocol. The schema is:

Schema 6: Circumspect team reasoning
(1) I am a member of T.
(2) It is T-conditional common knowledge that each member of T identifies with S.

(3) It is T-conditional common knowledge that each member of T wants the value of U to be maximised.
(4) It is T-conditional common knowledge that P uniquely maximises U, given the actions of non-members of T.

I should choose my component of P.

Using Schemata 4, 5 and 6, we now review some salient features of Bacharach's theory of team reasoning. We focus particularly on Schema 4, since this is the most transparent of the three, and most of its main features are also present in the other schemata.

Notice that Schema 4 refers to a *particular* profile A which maximises the value of the payoff function U. Premise (4) states that, within the group of team reasoners, it is common knowledge that A maximises U. In his notes for the unwritten chapter VIII, Bacharach is much concerned with the question of whether this premise is unnecessarily strong. He considers a variant schema of the form:[4]

Schema 7: Basic team reasoning
(1) I am a member of S.
(2) It is common knowledge in S that each member of S identifies with S.
(3) It is common knowledge in S that each member of S wants the value of U to be maximised.
(4) It is common knowledge in S that each member of S knows his component of the profile that uniquely maximises U.

I should choose my component of the profile that uniquely maximises U.

Bacharach notes to himself the 'hunch' that this schema is 'the basic rational capacity' which leads to *high* in Hi-Lo, and that it 'seems to be indispensable if a group is ever to choose the best plan in the most ordinary organizational circumstances'.

Notice that Schema 7 does not require that the individual who uses it know *everyone's* component of the profile that maximises U. In this sense, this schema does not represent the whole of what Bacharach understands as 'team reasoning'. Recall from section 2 of chapter 4 that 'someone "team-reasons" if she works out the best feasible combination of actions for all the members of her team, then does her part in it'. In contrast, premise (4) of Schema 7 states only that it is common knowledge in S that each member of S knows *his own* component of the profile that uniquely maximises U. Of course, one way in which this premise can be true is that the nature of the game being played is common knowledge in S, and that each individual, reasoning independently, works out which profile is best for the whole group, thus discovering his own component of that profile. But here is another way. Suppose *S* is a football team which will play an important game tomorrow. Every player in the team wants to maximise the probability that the team wins, and this is common knowledge.[5] All the players turn up at a meeting at which the coach proposes tactics for the game. However, each player listens only to those tactics that concern him. The goalkeeper does not listen to the tactics for taking penalties; the centre-forward does not listen to the goalkeeping tactics. Still, it is common knowledge that each player knows his own component of the team's tactics. Thus, if the players have sufficient confidence in the coach's tactical sense, premise (4) is true. Cases of this kind are important for Bacharach's analysis of communication, which we explain in section 5.

Returning to Schema 4, notice that its conclusion is an *unconditional* proposition of the form 'I should choose my component of the best profile'. The unconditional nature of this conclusion is crucial in the resolution of the problem posed by the Hi-Lo game. In contrast, the best-reply reasoning of classical game theory leads to conclusions about what one agent should do *conditional on what other agents can be expected to do*. Thus, in the Hi-Lo game, best-reply reasoning leads only to the conclusion 'If I expect my opponent to choose her component of the best profile, I should choose mine', and so to an infinite regress. But even though Schema 4 yields an unconditional conclusion, the only situations in which it tells the individual to choose his component in the best profile are ones in which it also tells the others to choose theirs. Further, these are always situations in which the other players identify with S: they are framing the decision problem as a problem 'for us'. Thus if each of them is rational, each will act on the conclusions of Schema 4, as applied to her case. And since it is common knowledge in *S* that everyone identifies with *S*, each player can work all this out. So whenever Schema 4 tells an individual to choose her component in the best profile, that individual has the *assurance* that the others (if rational) will choose theirs too. To put this another way: when the individual chooses her component in the best profile, she can construe this component as her part of a collective action that is actually taking place.

This combination of unconditionality and assurance does not carry over to Bacharach's theories of restricted and circumspect team reasoning, as represented in Schemata 5 and 6. In these theories, each member of T—the group that Bacharach calls the *team*—identifies with *S*. Thus each member of *T* wants the value of *U* to be maximised, where *U* represents what people want *as members of S*. Each member of *T* is told to do his part in a joint action by *T* to maximise *U*, given the behaviour of nonmembers of T; he can be assured only that the other members of T will do their parts. In other words: there is a joint action by the members of T to achieve an objective that 'belongs to' S. At first sight, one might think that this is an inessential feature of Bacharach's analysis: why can't we simply substitute 'identifies with T' for 'identifies with S' in premise (2) of each of Schemata 5 and 6? But Bacharach's theory of framing commits him to these premises as we have written them.

His hypothesis is that group identification is an *individual's* psychological response to the stimulus of a particular decision situation. It is not in itself a group action. (To treat it as a group action would, in Bacharach's framework, lead to an infinite regress.) In the theory of circumspect team reasoning, the parameter ω is interpreted as a property of a psychological mechanism—the probability that a person who confronts the relevant stimulus will respond by framing the situation as a problem 'for us'. The idea is that, in coming to frame the situation as a problem 'for us', an individual also gains some sense of how likely it is that another individual would frame it in the same way; in this way, the value of ω becomes common knowledge among those who use this frame. (Compare the case of the large cube in the game of Large and Small Cubes, discussed in section 4 of the introduction.) Given this model, it seems that the 'us' in terms of which the problem is framed must be determined by how the decision situation first appears to each individual. Thus, except in the special case in which $\omega = 1$, we must distinguish S (the group with which individuals are liable to identify, given the nature of the decision situation) from T (the set of individuals who in fact identify with S).

Finally, we consider the implications of Bacharach's claim that the reasoning in the three schemata is valid. Take Schema 4. Notice that the reasoning is *instrumental*. The conclusion—a proposition about what the individual should do—is derived from a set of premises which, among other things, state what the individual wants to achieve, namely that U is maximised. While asserting that Schema 4 is formally valid, Bacharach makes no claims about what, either rationally or morally, the individual *ought* to want to achieve. Similarly, that the individual identifies with S is an implication of premise (2). Bacharach makes no claims about which group the individual *ought* to identify with, or about whether the individual ought to identify with any group larger than the singleton that contains only herself. Schema 4 is presented as valid instrumental reasoning, *given* the unit of agency and *given* what the agent wants to achieve.

Because Bacharach uses an instrumental conception of rationality, he can assert the validity of both the team reasoning of Schemata 3–6 and the individual reasoning of Schema 1. Whether the individual reasons as an individual or as a member of some larger group and, if the latter, which larger group she reasons as a member of are matters of psychological framing, not rationality. This feature of Bacharach's approach allows for a great deal of flexibility, while retaining much of the formal structure of classical game theory. For a given game, there may be many different 'groups' of players. These may be larger or smaller, disjoint or overlapping. Some players may identify with one group, some with another; which group a given player identifies with may be determined by a random process. There can be strategic interaction between groups as well as between individuals. (For a discussion of these possibilities, see Bacharach 1999.) What it does not allow is that an individual might simultaneously feel a sense of identity with two or more different groups.

This might seem to be a limitation of Bacharach's theory: intuitively, one might think, there are situations in which individuals have to choose between conflicting group loyalties or identities. But in modelling the choices that people make, Bacharach's aim is to represent modes of practical reasoning by which those choices can be justified as rational. If rationality is conceptualised instrumentally, the idea of rational choice between alternative specifications of one's sense of agency, or between alternative specifications of one's fundamental objectives, is a contradiction in terms. To develop a theory of team reasoning which could encompass choice between identities would require a radical change in Bacharach's modelling strategy.

2. Commitment, Assurance and Rationality

In this section we draw out some significant features of Bacharach's theory through comparisons with some of the other extant theories of team reasoning and collective intentions. Bacharach presents a brief account of the connection between reasoning and intentions in chapter 4, section 5. On his account, an intention is interposed between a decision and an action, so it is natural to treat the intentions that result from team reasoning as collective intentions.

Bacharach's theory of team reasoning has strong similarities with Donald Regan's (1980) account of cooperative utilitarianism, and with Sugden's (1993) theory of 'thinking as a team'. As Bacharach points out, his theory of restricted team reasoning is formally similar to Regan's analysis, in which each member of the team of cooperative utilitarians identifies with the set of all people (or all sentient beings). As he also points out, the main differences between his theory and

Sugden's are concerned with restricted and circumspect team reasoning and with the related issue of assurance. Sugden presents assurance as a fundamental component of team reasoning. He suggests that his theory of team reasoning is in the same spirit as his earlier model of voluntary contributions to public goods, which is based on a moral principle of reciprocity (Sugden 1984). If team reasoning is to be interpreted in terms of reciprocity, assurance seems to be essential. In contrast, reciprocity does not seem to be central to Bacharach's understanding of team reasoning.

There is a related difference between, on the one hand, Bacharach's theories of restricted and circumspect team reasoning and, on the other, accounts of group agency in which a group is constituted by public acts of promising, or by public expressions of commitment by its members. This latter idea is central to Margaret Gilbert's (1989) analysis of 'plural subjects'. Gilbert is less concerned than Bacharach with how a plural subject should choose from a menu of options; her main concern is with what it means for a plural subject to *do* something—whether performing an action, holding a belief, or taking an attitude. Still, her analysis of how a plural subject is formed might be applied to the formation of teams in Bacharach's sense. There are hints of this approach in the work of Martin Hollis (1998). Hollis suggests that Jean-Jacques Rousseau's (1762/1988) account of the social contract, in which the act of forming a society brings about a 'most remarkable change in man', can be understood as a transition from individual to group agency that takes place through a collective act of commitment. It seems natural to think of joint commitment as involving some kind of reciprocity: in committing oneself to act as a member of a group, one looks for assurance that other members will be similarly committed.

However, Bacharach's and Gilbert's understandings of group identity may be more similar than their formal analyses suggest. In his notes for the unwritten chapter VIII, Bacharach records the following train of thought about the formation of teams:

> My current . . . idea is something like this: Something in the situation prompts the parties to see that they have action possibilities which provide joint agency possibilities which have possible outcomes of common interest. Each finds herself in a frame which features concepts which describe the conceived possible actions, describe the conceived possible outcomes, and present some of these outcomes positively. Each of us sees that we could write a paper together, or have a pleasant walk round the garden together, or bring down the appalling government together.

> The holism of frames comes in here. One concept belonging to the frame may bring others with it, or only be activated if others are. Some actions only get conceived if one gets the idea of certain possible outcomes, and conversely. One such situation is that created by one of us being so prompted, then making a verbal suggestion to the other(s), as in 'Would you like to dance?'

There are strong similarities here with Gilbert's account of the formation of plural subjects, except that what Gilbert describes in the language of agreement and tacit understanding, Bacharach describes in terms of framing. Bacharach's concept of framing allows one person to choose an action with the intention of affecting *someone else's* frame. While Gilbert would treat the saying of the words 'Would you like to dance?' as the first stage in a process that may lead to the two people's jointly committing themselves to acting as a plural subject, Bacharach treats it as part of a process by which two people influence each other's frames. Bacharach cannot say (as Gilbert might) that each individual *chooses* (or *agrees*, or *commits himself*) to view the situation in the 'we' frame. But he can, and does, say that group identification tends to be primed by an individual's recognition that the members of the putative group have common interests that can be furthered by joint action. So although Bacharach's individuals cannot choose to create teams with the rational intention of arriving at mutually beneficial solutions to problems of coordination or cooperation, the opportunities for mutual benefit offered by such problems tend to induce the kind of mutual adjustment of frames that he describes.

In Bacharach's analysis, team reasoning is instrumental. This coheres with his analysis of group identification as framing. However, if instead one follows Gilbert and thinks of group identification as the product of joint commitment, it may seem more natural to think that the rationality of acting as a member of a team derives from the rationality of fulfilling one's commitments or intentions. This seems to be Gilbert's position. Focusing on collective attitudes rather than on collective actions, she claims that membership of a plural subject imposes obligations to uphold 'our' attitudes. This claim is conceptual rather than moral: roughly, the idea is that a plural subject is formed by the making of joint commitments, and that each party to a joint commitment has the standing to demand that the others uphold their parts of that commitment.

Some contributors to the theory of team reasoning have wanted to go beyond Bacharach's instrumentalism by claiming that, in particular circumstances, rationality or morality requires individuals to *form* teams. For example, Regan commends cooperative utilitarianism as a rational morality. The fundamental principle of this morality is that 'what each agent ought to do is to co-operate, with whoever else is co-operating, in the production of the best consequences possible given the behaviour of non-co-operators' (1980, p. 124). This principle can be represented as an instance of Schema 5, in which S is the set of all persons and the value of U is a utilitarian measure of overall goodness. But then Regan's claim is not merely that this schema is valid. He also claims that every reasoning agent *ought* to identify with S, and *ought* to want U to be maximised. In a similar vein, Susan Hurley (1989, pp. 136–159) proposes that we (as rational and moral agents) should specify agent-neutral goals—that is, goals of which it can simply be said that they ought to be pursued, rather than that they ought to be pursued by some particular agent. Then we should 'survey the units of agency that are possible in the circumstances at hand and ask *what the unit of agency, among those possible, should be*'; and we should 'ask ourselves *how we can contribute to the realization of the best unit possible in the circumstances*'. The idea seems to be that a person may have the capacity to choose the unit of agency in which she participates, and that such choices should be governed by goals that are agent-neutral.

In contrast, Bacharach's theory makes group identification a form of framing. For Bacharach, frames are not chosen: they are individuals' psychological perceptions of the world. Recall his criticism of Gauthier's hypothesis that, in coordination games, players rationally choose which frame to use: '[It] seems to me that one can't just go round changing one's own description for convenience; this is like changing beliefs; surely you must describe the world as you find it!' (quoted in section 5 of the introduction). It seems that Bacharach is committed to a modelling strategy in which the unit of agency cannot be chosen. Nor can his modelling strategy admit the concept of a goal that is not the goal of some agent. Such goals cannot be represented within the instrumental conception of rationality that is fundamental to Bacharach's analysis of team reasoning.

Other contributors have wanted to give *less* prominence to rationality than Bacharach does. Sugden (2003) presents a 'logic of team reasoning' without making any claims for its validity. In Sugden's analysis, a 'logic' is merely an internally consistent system of axioms and inference rules. An individual actor may *endorse* a particular logic, thereby accepting as true any conclusions that can be derived within it, but the theorist need not take any position about whether the axioms of that logic are 'really' true or whether its inference rules are 'really' valid. Team reasoning is then represented as a particular inference rule which, as a matter of empirical fact, many people endorse. Thus, following this approach, one might reinterpret Schema 4 as specifying the inference rule 'From (1), (2), (3) and (4), infer "I should choose my component of A"'. On this interpretation, however, the premises of Schema 4 are too weak to guarantee the assurance that, for Sugden, is an essential part of team reasoning. He maintains assurance by requiring, as one of the premises from which 'I should choose my component' is inferred, that each member of S has reason to believe that each other member endorses and acts on team reasoning (and that each has reason to believe that each other has reason to believe that each other endorses and acts on team reason, and so on). Sugden shows that, despite first appearances, this analysis is not circular, and does not reduce to a form of contract between individually rational agents.

Our hunch is that Bacharach would not have sympathised with this approach. (We cannot know, since Sugden's logic of team reasoning was proposed only after Bacharach's death.) For Bacharach, it was fundamental that team reasoning should be represented as *valid* reasoning, and not merely as an algorithm or heuristic that people happen to use. And—as his theories of restricted and circumspect team reasoning show—he did not see assurance as essential to team reasoning.

3. The Third Puzzle of Game Theory: The Prisoner's Dilemma

Bacharach's analysis of the Prisoner's Dilemma is a straightforward application of his theory of team reasoning. It was presented in outline form in section 3.3 of chapter 4; we now expand on that discussion. Figure C.1 presents the Prisoner's Dilemma in a more general form. Instead of postulating specific numerical payoffs, we merely require that the payoffs be symmetrical between the players and that they satisfy two inequalities. The inequality $a > b > c > d$ encapsulates the central features of the Prisoner's Dilemma: that, for each player, the best outcome is that in which he chooses *defect* and his opponent chooses *cooperate*; the second-ranked outcome is that in which both

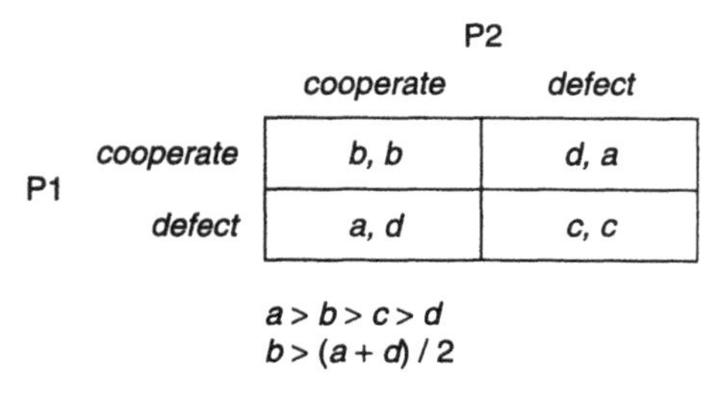

		P2 cooperate	P2 defect
P1	cooperate	b, b	d, a
P1	defect	a, d	c, c

$a > b > c > d$
$b > (a + d) / 2$

Figure C.1. The Prisoner's Dilemma

choose *cooperate;* the third-ranked is that in which both choose *defect;* and the worst is that in which he chooses *cooperate* and his opponent chooses *defect.* The inequality $b > (a + d)/2$ stipulates that each player prefers a situation in which both players choose *cooperate* to one in which one of them chooses *cooperate* and the other chooses *defect,* each player being equally likely to be the free-rider. This condition is usually treated as a defining feature of the Prisoner's Dilemma.

In the context of this game, it is natural to define *S*, the group with which a team reasoner identifies, as {P1,P2}. In order to apply Bacharach's theory of team reasoning, we need to define a payoff function U to represent what each individual wants to achieve, given that she identifies with {P1,P2}. Bacharach assumes that, when a player identifies with this group, she wants to promote the combined interests of its two members, at least insofar as those interests are affected by the game that is being played. Thus the values of U can be interpreted as measures of the welfare of the group {P1,P2}. If we assume that U treats the players symmetrically, we need to specify only three values of this function: the payoff when both players choose *cooperate,* which we denote u_C, the payoff when both choose *defect,* which we denote u_D, and the payoff when one chooses *cooperate* and one chooses *defect,* which we denote u_F (for 'free-riding'). It seems unexceptionable to assume that U is increasing in individual payoffs, which implies $u_C > u_D$. (This is the 'Paretianness' condition that Bacharach defends in section 4.3 of chapter 2.) Given the condition $b > (a + d)/2$, it is natural also to assume $u_C > u_F$. Then the profile of actions by P1 and P2 which uniquely maximises *U* is (*cooperate, cooperate*).

How the players are predicted to act depends on what is assumed about framing. The Prisoner's Dilemma has the property of 'strong interdependence' (defined in section 4.1 of chapter 2). Recall that, although Bacharach proposes that the perception of this property increases the probability of group identification, he does not claim that the Prisoner's Dilemma *invariably* primes the 'we' frame:

> In a Prisoner's Dilemma, players might see only, or most powerfully, the features of common interest and reciprocal dependence which lie in the payoffs on the main diagonal. But they might see the problem in other ways. For example, someone might be struck by the thought that her coplayer is in a position to double-cross her by playing [*defect*] in the expectation that she will play [*cooperate*]. This perceived feature might inhibit group identification. (chapter 2, section 4.2)

The implication is that the 'we' frame *might* be primed; but, alternatively, a player may see the game as one to be played by two separate individual agents.

That either framing is psychologically possible reflects the sense in which the Prisoner's Dilemma is one of the puzzles of game theory. On the one hand, the positions of the two players are completely symmetrical, which prompts each player to focus on strategy profiles in which the players' actions are symmetrical. Then, comparing the outcomes of (*cooperate, cooperate*) and (*defect, defect*), each sees that the two players have a common interest in their both choosing *cooperate*. This line of thought leads naturally to a conception of the game as a problem 'for us'. On the other hand, if one looks at the outcomes of (*cooperate, defect*) and (*defect, cooperate*), one sees a conflict of interest between the two players: by choosing *defect* when one's opponent chooses *cooperate*, one can gain at her expense. This latter line of thought leads to a conception of the game as one in which the two players are in opposition, each facing her own decision problem. As a metaphor or model, Bacharach sometimes refers to the famous drawing, used in Gestalt psychology and reproduced as figure C.2, which can be seen either as a vase or as two faces. In the same way, the Prisoner's Dilemma can be seen by a player either as a problem 'for me' or as a problem 'for us'.

Figure C.2. The vase–faces illusion

In Bacharach's theoretical framework, this dualism is best represented in terms of circumspect team reasoning. Let ω (where $1 \geqslant \omega > 0$) be the probability that, for any individual player of the Prisoner's Dilemma, the 'we' frame comes to mind; if it does, the player identifies with {P1,P2}. Assume that, if this frame does *not* come to mind, the player conceives of himself as a unit of agency and thus, using best-reply reasoning, chooses the dominant strategy *defect*. We can now ask which protocol maximises U, given the value of ω. Viewed from within the 'we' frame, the protocol (*defect, defect*) gives a payoff of u_D with certainty. Each of the protocols (*cooperate, defect*) and (*defect, cooperate*) gives an expected payoff of $\omega u_F + (1 - \omega)u_D$. The protocol (*cooperate, cooperate*) gives an expected payoff of $\omega^2 u_C + 2\omega(1 - \omega)u_F + (1 - \omega)^2 u_D$. There are two possible cases to consider. If $u_F \geqslant u_D$, then (*cooperate, cooperate*) is the U-maximising protocol for all possible values of ω. Alternatively, if $u_D > u_F$, which protocol maximises U depends on the value of ω. At high values of ω, (*cooperate, cooperate*) is uniquely optimal; at low values, the uniquely optimal protocol is (*defect, defect*).[6]

If we assume *either* that $u_F \geqslant u_D$ *or* that the value of ω is high enough to make (*cooperate, cooperate*) the uniquely optimal protocol, we have a model in which players of the Prisoner's Dilemma choose *cooperate* if the 'we' frame comes to mind, and *defect* otherwise. Bacharach offers this result as an explanation of the observation that, in one-shot Prisoner's Dilemmas played under experimental conditions, *cooperate* and *defect* are each usually chosen by a substantial proportion of players. He also sees this result as consistent with the fact that there are many people who think it completely obvious that *cooperate* is the only rational choice, while there are also many who feel the same about *defect*. Bacharach can say that both sets of people are right—in the same way that two people can both be right when one says that the drawing they have been shown is a picture of a vase and the other says it is a picture of two faces.[7]

4. Other Theories of Cooperation

Bacharach aligns himself with the small minority of theorists who maintain that a rational player of the one-shot Prisoner's Dilemma can choose *cooperate*. For many game theorists, this conclusion is close to heresy. For example, Ken Binmore (1994, pp. 102–117, quotation from p. 114) argues that it can be reached only by 'a wrong analysis of the wrong game': if two players truly face the game shown in Figure C.1, then it follows from the meaning of 'payoff' and from an unexceptionable concept of rationality that a rational player must choose *defect*. His argument works as follows. Consider player P1. She knows that her opponent must choose either *cooperate* or *defect*. The inequality $a > b$ tells us that, if P1 knew that P2 would choose *cooperate*, P1 would want to choose, and would choose, *defect*. The inequality $c > d$ tells us that, if P1 knew that P2 would choose *defect*, P1 would want to choose, and would choose, *defect*. So (Binmore concludes) we need only a principle of dominance to conclude that, whatever P1 believes about what P2 will do, P1 should choose *defect*. Binmore recognises that rational individuals may sometimes choose *cooperate* in games in which *material payoffs*—that is, outcomes described in terms of units of commodities which people normally prefer to have more of rather than less, such as money, or years of not being in prison—are as in figure C.1. But that just shows that the payoffs that are relevant for game theory—the payoffs that govern behaviour—can differ from the material ones. The first stage in a game-theoretic analysis of a real-life situation should be to find a formal game that correctly represents that situation.

Thus in response to the problem of explaining why *cooperate* is sometimes chosen in games whose material payoffs have the Prisoner's Dilemma structure, the methodological strategy advocated by Binmore is that of *payoff transformation*: we should look for some way of transforming material payoffs into game-theoretic ones which makes observed behaviour consistent with conventional game-theoretic analysis. This strategy has been followed by various theorists who have proposed transformations of material payoffs to take account of psychological or moral motivations that go beyond simple self-interest. For example, in slightly different ways, Ernst Fehr and Klaus Schmidt (1999) and Gary Bolton and Axel Ockenfels (2000) propose that, for any given level of material payoff for any individual, that individual dislikes being either better off or worse off than other people. Matthew Rabin (1993) proposes that each individual likes to benefit people who act with the intention of benefiting him, and likes to harm people who act with the intention of harming him.[8]

Clearly Bacharach cannot accept Binmore's argument. But what is wrong with it? Bacharach accepts Binmore's instrumental conception of rationality, but rejects his implicit assumption that agency is necessarily vested in individuals. Bacharach interprets the payoffs of a game, as represented in a matrix like that of figure C.1, as showing what each player wants to achieve *if she takes herself to be an individual agent*. In this sense, Bacharach's position is close to Binmore's: payoffs are defined, not in material terms, but in terms of what individuals are seeking to achieve. Bacharach endorses Binmore's analysis when it is applied to players who take themselves to be individual agents. Thus Bacharach agrees that if P1 frames the game as a problem 'for me', the only rational choice is *defect*. However, he also allows the possibility that P1 frames the game as a problem 'for us'. In that case, the payoffs that are relevant in determining what it is rational for P1 to do are measures of what P1 wants to achieve *as a member of the group {P1,P2}*; and these need not be the same as the payoffs in the standard description of the game.

Thus there is a sense in which Bacharach's explanation of the choice of *cooperate* in the Prisoner's Dilemma depends on a transformation of payoffs from those shown in figure C.1. However, the kind of transformation used by Bacharach is quite different from that used by theorists such as Fehr and Schmidt. Bacharach's transformation is not from material payoffs to choice-governing payoffs; it is from payoffs which govern choices for one unit of agency to payoffs which govern choices for another.

One might wonder whether Bacharach needs to transform *both* payoffs *and* agency. If payoffs have been transformed so that they represent the welfare of the two players as a group, doesn't conventional game theory provide an explanation of why each individual chooses *cooperate*? Not necessarily. Consider a Prisoner's Dilemma in which $a = 10$, $b = 8$, $c = 6$ and $d = 0$, and assume that the value of the payoff function for the group {P1,P2} is the average of the payoffs for the two individuals. Then we have $u_C = 8$, $u_D = 6$ and $u_F = 5$. If we treat P1 and P2 as individual agents, each of whom independently seeks to maximise the value of U, we have the game shown in figure C.3. The structure of this game will be familiar: it is yet another variant of Hi-Lo, in which *cooperate* corresponds with *high* and *defect* with *low*. Conventional game theory does not show that rational players of this game will choose *cooperate*. To show that, we need a transformation of the unit of agency.

By using the concept of agency transformation, Bacharach's theory is able to explain the choice of *high* in Hi-Lo. Existing theories of payoff transformation cannot do this. It is hard to see how *any* such theory could credibly make (*high, high*) the unique solution of Hi-Lo. Let us interpret the Hi-Lo payoffs as material payoffs, and consider possible transformations. Suppose that, following Fehr and Schmidt and Bolton and Ockenfels, we introduce assumptions about players' attitudes towards the distribution of material payoffs. In every possible outcome,

		Player 2	
		cooperate	*defect*
Player 1	*cooperate*	8, 8	5, 5
	defect	5, 5	6, 6

Figure C.3. A Prisoner's Dilemma with transformed payoffs

the two players' material payoffs are equal. So whatever the players' attitudes to inequality, it seems that their subjective ranking of the outcomes must correspond with the ranking of material payoffs: a game that is Hi-Lo in material payoffs will remain Hi-Lo after payoff transformation. Alternatively, suppose we follow Rabin and assume that each player wants to reciprocate other players' 'kindness' or 'unkindness' towards him. In a situation in which each of the players P1 and P2 chooses *low*, P1 is benefiting P2 to the maximum degree possible, given P2's action, and vice versa. So each is reciprocating the other's 'kindness'. Reciprocity in Rabin's sense does not affect the equilbrium status of (*low*, *low*).

For Bacharach, the 'strongest argument of all' in support of his theory of cooperation is that the same theory predicts the choice of *high* in Hi-Lo games. He claims that any satisfactory theory of rational choice should explain why *high* is chosen. Once we have the components that Bacharach's theory needs in order to arrive at this explanation, very little more is needed to explain the choice of *cooperate* in the Prisoner's Dilemma: 'A way of framing games that is necessary for people to conform to the innocuous [payoff dominance principle] is also sufficient for cooperating in Prisoner's Dilemmas'.[9]

However, Bacharach recognises that his is not the only theoretical approach that can be used to explain both types of behaviour as rational. One such approach is that of theories of *conditional cooperation*, which postulate the existence of a *self-recognising* set *X* of individuals—that is, a set with the property that each member has the ability to recognise the others—who cooperate with one another. In the case of the Prisoner's Dilemma, the assumption is that members of *X* choose *cooperate* if playing against comembers, but *defect* otherwise. It is easy to see that, if members of *X* meet one another with positive probability and if nonmembers of *X* play *defect* unconditionally, the expected payoff for a member of *X* is higher than that for a nonmember. Thus if membership of *X* is treated as a matter of choice, it is instrumentally rational to become a member. A similar argument can be used in relation to Hi-Lo. In a population of players of Hi-Lo in which almost everyone chooses *low*, the members of a self-recognising set *X* of individuals who play *high* against one another and *low* against non-members will do better than non-members. (The converse does not hold: in a population in which almost everyone chooses *high*, it is not advantageous to join a group whose members play *low* against one another. Thus self-recognition induces the choice of *high*.)

A particularly famous example of this theoretical genre is David Gauthier's (1986) theory of *constrained maximisation*. In Gauthier's theory, the members of *X* are defined by their sharing a certain 'disposition' which leads them to cooperate with comembers. This disposition is assumed to be sufficiently 'translucent' that comembers can usually recognise one another. Gauthier analyses choice among dispositions in terms of instrumental rationality: each individual chooses a disposition with the aim of maximising his expected payoff as an individual. Gauthier's theory is presented in relation to the Prisoner's Dilemma, but it might be extended to apply to Hi-Lo.

The theories of constrained maximisation and team reasoning have some affinities. Each of these theories treats game-theoretic payoffs as measures of what individuals want to achieve (rather than as material payoffs in need of transformation). Each uses some kind of maximising criterion to determine each individual's rational or moral action. Each differs from conventional game theory by viewing the individual's decision problem in some nonstandard perspective. In Gauthier's theory, the change in perspective is from choice among actions to choice among dispositions; in Bacharach's, it is from individual agency to team agency. There is a fundamental difference, however, between the ways in which the two theorists conceive of what we have called 'changes in perspective'. In Bacharach's theory, the unit of agency is determined by a psychological mechanism of framing; the instrumental conception of rationality that underlies his analysis can be applied only *within* a particular frame. Thus Bacharach does not claim that it is instrumentally rational for an individual to adopt the 'we' frame. In contrast, Gauthier argues that it is instrumentally rational for each individual to construe decision problems in terms of choice among dispositions.[10] This argument depends critically on the assumption that dispositions—understood not as psychological propensities but as attitudes that can be taken on by acts of conscious choice—are translucent to others. Many commentators have thought that assumption implausible.[11] In effect, Gauthier's analysis makes cooperative behaviour a special case of individually instrumental choice. For Bacharach, this feature of Gauthier's theory disqualifies it from being a 'true cooperative theory': a true theory of cooperation must explain acts that are carried out with cooperative *intent*.[12]

Another approach to explaining cooperation, discussed in section 6.1 of chapter 1, assumes that agents engage in *magical thinking*. Suppose that P1 and P2 are involved in some strategic interaction. Suppose that, in P1's framing of this interaction, the set of act-descriptions from which she chooses is the same as the set from which P2 chooses, and that the payoff matrix is symmetrical with respect to the players. P1 engages in magical thinking if she reasons from the premise that, whichever action she chooses, P2 will make the same choice. In the case of the Prisoner's Dilemma, P1 reasons that if she chooses *cooperate*, so will P2, while if she chooses *defect*, so will P2. Since the outcome of (*cooperate*, *cooperate*) is better for P1 than the outcome of (*defect*, *defect*), P1 should choose *cooperate*. Intuitively, this reasoning is supported by the thought that P1 and P2 both face the same problem, and that (at least if they are sufficiently similar in psychology, experience and so on) are likely to think about it in similar ways. Thus the fact that P1 chooses, say, *defect* seems to be evidence that P2 is likely to choose *defect* too. Clearly the same kind of reasoning would recommend *high* in Hi-Lo.

Most decision theorists view such reasoning as fallacious: when a person is reasoning about what to choose, she should take account of the *causal* consequences of her actions, but not of what those actions are *evidence* of. Bacharach agrees: he treats magical thinking as one of a class of 'modes of reasoning . . . which seem clearly fallacious' and which are 'dysfunctional in important ranges of problems'.[13] Since his concern is with *valid* reasoning, he has no reason to pursue this approach.

The term 'magical thinking' is used to suggest that the person who reasons in this way believes that her own actions have some mysterious causal power over other people's. However, there are modes of reasoning which, at first sight, can seem like magical thinking but which are not fallacious. In everyday moral argument, it is a standard move to ask 'How would you like it if everyone behaved like that?' The suggestion is that, in deciding how to act, each individual ought to reason *as if* her chosen action would also be chosen by anyone else in the same circumstances. This is not an instrumental 'ought', resting on the false assumption that each individual's action determines the actions of the others. Rather, it is a moral 'ought'. Some theorists have proposed *deontological* explanations of cooperation which assume that individuals recognise moral requirements of this kind. In these theories, the conventional assumption of instrumental rationality is supplemented by the assumption that individuals accept certain moral constraints on the pursuit of self-interest.

Several writers have proposed deontological theories in which each individual considers alternative behavioural rules, each of which is formulated so as to apply to everyone in some group; each individual is morally obliged to act on the rule that he would most prefer everyone in that group to follow. The rule picked out by this criterion is often called the *Kantian* rule. Theories of this kind have been proposed by Jean-Jacques Laffont (1975), David Collard (1978) and John Harsanyi (1980). Sugden (1984) proposes a related theory of *reciprocity* in which each individual is required to follow the Kantian rule provided that she has reason to believe that everyone else will follow it.

In the context of the Prisoner's Dilemma, the Kantian rule is 'Choose *cooperate*'. In the context of Hi-Lo, it is 'Choose *high*'.

These deontological theories can be thought of as occupying a space between individual rationality and team reasoning. In deciding what to do, the Kantian agent treats his own action as part of a hypothetical profile of similar actions by everyone, in much the same way that a team reasoner does. But, while the team reasoner evaluates the outcome of such a profile in terms of what he takes to be the good of the group, the Kantian evaluates it in terms of what he wants to achieve as an individual. As far as we know, Bacharach never compares team-reasoning theories with Kantian ones. Our guess is that he would have argued that the team-reasoning approach is intuitively more coherent, and that it is compatible (as the Kantian approach is not) with the instrumental conception of rationality that is fundamental to classical decision and game theory.

5. Communication

So far, the theory of team reasoning has been presented, both by Bacharach and by us, in relation to games in which there is no communication between players and in which players' actions are chosen simultaneously. Given that the object is to explain coordination and cooperation, focusing on games with these properties may seem artificial. Most real-world coordination and cooperation, and most of what is naturally called team activity, involves communication or sequential moves. These two features of games are closely connected. We might model communication by starting with an *underlying game* (such as the Prisoner's Dilemma) in which players act simultaneously, and then assuming that, before that game is played, the players have opportunities to send messages to each other. But if we model those opportunities formally, we have an expanded game in which acts of communication are moves, and not all moves are simultaneous. Conversely, in a game with sequential moves, moves can communicate information about players' preferences or intentions.

Bacharach's notes for chapter VIII show that he was very conscious of this apparent objection. He intended to extend his analysis to problems of coordination and cooperation in situations which allow communication and in which individuals act sequentially. He calls such situations *organisations* (on the grounds that 'the essence of organization is information distribution').[14] It is clear from these notes that Bacharach thinks he can show that coordination and cooperation in organisations depend on essentially the same kind of team reasoning as applies in his analysis of the three 'puzzles': pure coordination games, the Prisoner's Dilemma and Hi-Lo. However, the analysis by which this is to be shown is incomplete. Those parts that have been set down are in the form of abbreviated notes. The whole structure of the argument was constantly being revised right up to Bacharach's death: the notes often read as if he is arguing with himself. In this section, which considers communication, and section 6, which considers sequential moves, we try to follow through these unfinished lines of argument.

In this section, we focus on two general organisational structures, each of which implies a variant form of each of the three types of game. The first of these is *direction*. In this structure, one of the players is picked out, prior to the game itself, to be the *director*. There is an underlying game in which the players choose actions simultaneously. Before those choices are made, the director has the opportunity to send a common message to all the other players. There are no other message-sending opportunities: the other players cannot reply to the director's message, or send messages to one another. Bacharach's example of calling runs in cricket (chapter 1, section 1) illustrates this organisational structure. The underlying game is a variant of Hi-Lo, in which each player chooses between the options *run* and *stay*. Depending on where the ball has been hit, one of the two batsmen assumes the role of director; the message is a call of 'Yes' (indicating *run*) or 'No' (indicating *stay*).

The second structure is *deciding together*. As in the case of direction, there is an underlying game with simultaneous moves. Before the underlying game is played, the players meet together for a period of open discussion. This discussion may (but need not) end with expressions of agreement about which action each player should take. Whatever words are spoken in the discussion period, these do not constrain the players' freedom of action in the underlying game. Bacharach gives this example:

> Suppose you and some others have a shared aim, to rob a bank and not get caught, and meet and work out a plan, which you agree is the best feasible one, and which specifies that you wait at the fifth lamppost in Chancery Lane from 4.35 a.m. and when approached by a stranger in a felt hat asking the time, reply 'Phoebus' bright chariot doth soar to heaven's peak' . . .

The suggestion is that, on the morning of the robbery, you and the others will be playing a Hi-Lo game in which there are a huge number of alternative profiles of actions, only a tiny proportion of which will lead to success. Because of the need for secrecy, there can be no communication during the robbery itself. The mechanism for coordinating actions is prior agreement on a plan.

We begin by focusing on Bacharach's three puzzles in their purest forms—the one-shot games of Heads and Tails, Hi-Lo and the Prisoner's Dilemma, as presented in the introduction. These are puzzles for game theory in its classical form, which assumes all of the following: that choices respect invariance (that is, they are independent of descriptions: see section 3 of the introduction); that each player of a game maximises expected utility, given the expected behaviour of other players; that players have common knowledge of the game itself; and that they have common knowledge of each other's rationality. Given only those assumptions, we cannot explain how players of Heads and Tails reason to the choice of *heads*, how players of Hi-Lo reason to *high*, or how players of the Prisoner's Dilemma (sometimes) reason to *cooperate*. The first question we need to ask is whether these negative conclusions are affected if the players of these games have opportunities for preplay communication. The answer is simple: the negative conclusions remain.

First, consider Heads and Tails. Without preplay communication, this is a game in which each player chooses one of two strategies. Given the invariance assumption, these strategies are completely symmetrical with each other, and so any reason for choosing *heads* is also a reason for choosing *tails*. Now suppose we embed this game in an organisational structure of direction. To keep things simple, suppose that player P1 is the director, and that he can choose one of two costless messages to send to P2: 'Choose *heads*' or 'Choose *tails*'. At first sight, it seems that this solves the problem. Surely, one might say, rationality requires both players to choose *heads* if the first message is sent and to choose *tails* if the second is sent. And then, whichever message P1 sends, coordination is secured. But this reasoning neglects invariance. Given invariance, all we can say is that P1 has a choice between two opening moves; that one is *called* 'Choose *heads*' and the other 'Choose *tails*' is just a matter of labelling. Since the payoff matrix in the underlying game is unaffected by which of the two opening moves P1 makes, those moves are symmetrical with each other from both players' points of view. Any reason for P1 to choose one message is also a reason for his choosing the other. Any reason for P1 to choose *heads* conditional on having sent the message 'Choose *heads*' is also a reason for him to choose *tails* conditional on having sent that message. And any reason for P2 to choose *heads* conditional on receiving the message 'Choose *heads*' is also a reason for her to choose *tails* conditional on receiving that message.

Of course, it is obvious to human players that the message 'Choose *heads*' singles out *heads* as the action to be chosen by each of P1 and P2 in the underlying game. But that is a property of labelling, not of classical game-theoretic rationality. The message works by giving *heads* a particular form of salience, namely that this is the action that has been referred to in the only message that has passed between the players. But we already know that, in the absence of preplay communication, *heads* is salient, by virtue of the conventional priority of 'heads' over 'tails'. The problem posed by Heads and Tails is not that the players lack a common understanding of salience; it is that game theory lacks an adequate explanation of how salience affects the decisions of rational players. All we gain by adding preplay communication to the model is the realisation that game theory also lacks an adequate explanation of how costless messages affect the decisions of rational players.

Essentially the same analysis applies to Hi-Lo and the Prisoner's Dilemma, and to deciding-together structures as well as to direction structures. Once we invoke the invariance principle, we find that every costless message is equivalent to every other. Thus no amount of transmission of costless messages can provide a player with reasons for choosing one action rather than another. Bacharach's three puzzles remain puzzles.

Nevertheless it is clear that, in reality, communication can supplement team reasoning in facilitating coordination in situations of uncertainty. Take the case of calling runs in cricket. In practice, two batsmen would enjoy less success in scoring runs if communication between them were forbidden (in the way that communication between partners is forbidden in bridge). They would still share a common objective, and they would still be able to use team reasoning; but, given the unpredictability of cricket, situations would be liable to arise in which they reached different conclusions about which of (*run, run*) and (*stay, stay*) had the higher expected payoff in terms of that common objective. Bacharach's chapter VIII would have presented an analysis of how communication interacts with team reasoning.

Bacharach's notes suggest that he was exploring two different analyses of communication, and had not finally decided which he preferred. What we will call the *information-transmission analysis* treats organisation as a mechanism for delegating decision-making within a team, and treats communication as a mechanism for transmitting information about which profile of actions maximises the team's objective. For example, in the case of calling runs, the conventional practice delegates the decision to the batsman with the better view of the relevant area of the playing field. That batsman (say, P1) reaches a judgement about whether (*run, run*) or (*stay, stay*) is better for the team; his call of 'Yes' or 'No' is interpreted by the other batsman (P2) as an expression of that judgement. P2 is not required to make any personal judgement on the matter at all. He simply accepts P1's judgement about which profile of actions is better for the team, and then uses team reasoning to arrive at the conclusion that he should choose his component of that profile.

One feature of the information-transmission analysis is that it does not require that each member of the team know the whole profile of actions that is best for the team. It is sufficient that it is common knowledge that each member knows his own component of that profile. Each player can then use the 'basic team reasoning' of Schema 7. As an example, here is how Bacharach continues the story of the meeting to plan the robbery:

> As the meeting breaks up, the planner says: 'So remember, Marcus, the fifth lamppost, 4.35 a.m., . . . 'and similarly with the others. 'Does everyone know what he has to do?' What gives you good reason to do your part? The answer that each expects the other to do [his] part is clearly question-begging. What would be the basis for your expectation? The others' reasons. . . . Your having worked things out together gives each of you a [Schema 7]-based reason—namely, common knowledge that the best plan specifies that you wait at the fifth lamppost etc, and specifies something for each other, which they will remember. So if [Schema 7] is a legitimate schema of practical reasoning, your having worked things out together justifies your respective intentions.

Notice that it is not part of the story that each member of the gang remembers everyone's part in the plan. It seems that the meeting has served two purposes. First, it has made it common knowledge that the proposed plan is the best one. That need not imply that each member of the gang is capable of reproducing the reasoning that shows the optimality of the plan. Perhaps everyone merely knows that the planner can be relied on to find the best plan. Or perhaps the less experienced members of the gang can rely on the optimality of a plan that has passed the scrutiny of the more experienced members. All we have been told is that all the members of the gang have *agreed*, presumably through expressions of assent in the meeting, that the proposed plan is the best. Second, the end of the meeting has been structured to make it common knowledge that each member knows, and can be relied on to remember, his part of the plan. In these ways, the meeting ensures that the premises of Schema 7 are satisfied.

However, the information-transmission analysis has the same fundamental limitation as classical game-theoretic analyses of communication: it does not explain how costless messages come to be interpreted as conveying the information that their words assert. Take the case of calling runs. What makes it rational for the nonstriking batsman to interpret the striker's call of 'Yes' as information that, in the striker's judgement, (*run, run*) is the best profile of actions for the team? The information-transmission analysis cannot answer such questions.

Bacharach's alternative analysis of organisation attempts to overcome this problem. For reasons that will become clear shortly, we call this the *oracle analysis*. In this analysis, an announcement that asserts that a particular profile is best for the team is treated as a move (by a player, or by 'nature') in a larger game. As a result of that move, the components of that profile acquire a particular label, namely that they are components of what, *according to the announcement*, is the best profile. But for each player, it remains an open question which profile really *is* best.

Bacharach was exploring this analysis through what he called *oracle games*. Here is a simple example.[15] Suppose that P1 and P2 have been attending a conference. It is now evening, and they would like to eat together. They had expected their paths to cross during the day, but in fact they did not; they now have no way of communicating with each other. It is common knowledge between P1 and P2 that there are four restaurants in the town, one in each of the town's *north, south, east* and *west* quarters, and that these are all equally good. However, it is also common knowledge between them that the conference programme contains the words: 'By far the best restaurant in town is the one in the *east*'. In this case, the programme plays the role of *oracle*: it generates a statement asserting that a particular profile of actions in the underlying game is best for the relevant team. Since, by assumption, it is common knowledge between P1 and P2 that this assertion is false, the oracle's words cannot activate Schema 7 in the way that the planner's words did in the example of the robbery.

Nevertheless, the existence of the oracle affects the players' framing of the underlying game. First, consider the game in the absence of the oracle. Suppose there are just two relevant families of predicates that can be applied to the players' actions, the generic family F_0 = {*thing*} and the location family F_1 = {*north, south, east, west*}; and suppose that both families have availability 1. The objective game is a pure coordination game: both players want to go to the same restaurant, but are indifferent about which restaurant they go to. Thus, after applying the principle of insufficient reason, there is only one admissible act-description for each player: *pick a thing*. Now let us introduce the oracle. We now have a new family of predicates F_2 = {*recommended*}.[16] If we assume that this family has availability 1, the framed game has two admissible act-descriptions for each player: *pick a thing* and *choose the recommended*. Clearly the best profile of act-descriptions for both players—and hence, given the Paretianness assumption, the best for the team which comprises them both—is (*choose the recommended, choose the recommended*). Thus team reasoning tells each player to act on the oracle's recommendation. The point of this example is that, although the oracle's recommendation in itself has no information content, its existence makes the policy of acting on it optimal for the team of players which has access to it. The recommendation benefits the players, not by providing them with information about the objective game, but by changing the frame in which that game is viewed.

Notice that, in order to benefit from this property of the oracle's message, the players have to use Schema 4 rather than Schema 7. That is, each player has to be able to reason independently to the conclusion that a particular profile in the framed game, namely (*choose the recommended, choose the recommended*), is optimal for the team. Bacharach's analysis of oracle games seems to have led him to question his previous 'hunch' that Schema 7 (which he calls 'T') is more fundamental than Schema 4 (which he calls 'TR', for 'team reasoning'). As part of his notes on oracle games, he wrote:

> After much agonizing that the supposed reducibility of T to TR is a fix, I seem to have found some further support for the primacy of TR. . . . The higher is the correct level of analysis. That is, we must analyse at a level at which there is common knowledge of the payoff matrix. At this level it is TR that is needed.

The oracle analysis provides some explanation of how, in an organisational structure of direction, a director's message can be interpreted by the other players as an expression of his judgement. That analysis shows that, in certain types of coordination game, it is in the interests of the team that each member act on a recommendation from the director, even if that recommendation has no information content. In such games, team reasoning tells the director to make *some* recommendation, if necessary by picking a 'recommended' profile of actions at random. But then, provided that the director's judgements have *some* validity—that is, provided they are a better guide to what really is good for the team than random picking—team reasoning also tells the director to recommend whichever profile he judges to be best for the team. So, if it is common knowledge that the director is rational and identifies with the team, it is also common knowledge that his recommendations express his judgements.

The oracle analysis shows that organisational structures of the direction or deciding-together kinds can assist team reasoning, even if the truth values of the messages on which team members act are questionable. For example, in a direction structure, it can sometimes be optimal for the members of a team to act on a director's judgements about which profile of actions in the underlying game is best for the team, even though it is common knowledge that the director is the *least* capable judge in the team.

To say this is not to endorse the kind of unreasoning obedience immortalised in Alfred Tennyson's account of the charge of the Light Brigade.[17] Let S be the group with which team reasoners identify, and to whom the director's messages are addressed. Suppose the director's message asserts that a specific profile of actions A is the best in terms of U, the payoff function that all members of S want to be maximised. But suppose also that there is some subset of S, namely T, within which it is common knowledge that the director is wrong, and that (given that nonmembers of T act on the director's misjudged instructions) the best profile of actions for the members of T is A_T. Then restricted team reasoning (that is, Schema 5) tells each member of T to choose his component of A_T, even if that is contrary to the director's message. Notice, however, that in order for this reasoning to be activated, it has to be common knowledge in T that some *specific* profile, namely A_T, maximises U. In many practical cases, if the director is the only person with the opportunity to send messages, such common knowledge may not exist. For example, it may be common knowledge among the inhabitants of a country that everyone would be better off if their repressive ruler were overthrown, but a revolution or coup may be impossible unless the would-be rebels coordinate on a specific plan of action. Disabling the organisational structures through which such coordination could take place is one of the ways in which repressive regimes perpetuate themselves. Conversely, such regimes are often at their weakest when some event (perhaps a military defeat, or a historically significant anniversary) provides a focal point for a display of disaffection.

6. Sequential Moves and Backward Induction

One of the most famous pieces of game-theoretic analysis in philosophy is David Hume's story of two farmers who fail to help one another:

> Your corn is ripe to-day; mine will be so to-morrow. 'Tis profitable for us both, that I shou'd labour with you to-day, and that you shou'd aid me to-morrow. I have no kindness for you, and know you have as little for me. I will not, therefore, take any pains upon your account; and should I labour with you upon my own account, in expectation of a return, I know I shou'd be disappointed, and that

> I shou'd in vain depend upon your gratitude. Here then I leave you to labour alone: You treat me in the same manner. The seasons change; and both of us lose our harvests for want of mutual confidence and security. (1739–40/1978, pp. 520–521)

This example can be interpreted as a variant of the Prisoner's Dilemma, in which P1 and P2 are the two farmers, *cooperate* is the action of helping the other farmer, and *defect* is the action of not helping. Each act of helping imposes a cost on the helper and confers a greater benefit on the person being helped; thus each prefers not to help the other unilaterally, but a practice of mutual aid would benefit both. The difference from the classic Prisoner's Dilemma is that the two players move sequentially. The first mover in the game (P1) is the farmer whose corn ripens second; he chooses *cooperate* or *defect*. Then, having observed P1's action, P2 chooses *cooperate* or *defect*.

Does this sequential structure change the standard conclusion that, in a Prisoner's Dilemma, the individually rational action for each player is *defect*? Hume clearly thinks not, and most game theorists would agree. The standard method of analysing finite-length games with sequential moves—for short, *sequential games*[18]—is *backward induction*. This method, which dates back to Ernst Zermelo's (1912) classic analysis of Chess, works by considering every *path* through the game—that is, every sequence of moves that makes up a complete play of the game. First, we look at the final move of each path, and ask whether it is optimal for the relevant player. We deem that move to be *permissible* if and only if it is optimal. Having deleted all paths whose last moves are impermissible, we look at the penultimate move of every surviving path and ask whether that move is optimal for the relevant player, given the assumption that, whatever is done at that stage, the final move will be a permissible one. And so on, right back to the opening move. A backward-induction analysis of Hume's game begins by considering P2's move. If P1 has chosen *cooperate*, *defect* is uniquely optimal for P2; so *cooperate* is an impermissible response to *cooperate*. Similarly, if P1 has chosen *defect*, *defect* is uniquely optimal for P2; so *cooperate* is an impermissible response to *defect*. Now we consider P1's move, on the assumption that *cooperate* will be followed by *defect* and that *defect* will be followed by *defect*. Given this assumption, *defect* is uniquely optimal for P1. Thus the only permissible path—that is, the only path all of whose moves are permissible—is (*defect*, *defect*).

Just as we can construct a sequential form of the Prisoner's Dilemma, so we can construct sequential forms of Heads and Tails and Hi-Lo. In each case, we simply specify that P1 moves first and that P2 moves second, having observed P1's move; the payoff matrix is just as in the simultaneous-move game. A backward-induction analysis of the sequential form of Heads and Tails shows that the uniquely optimal move for P2 is to match whatever P1 has chosen—to respond to *heads* by *heads* and to *tails* by *tails*. Thus, *heads* and *tails* are equally good choices for P1. There are two permissible paths, (*heads, heads*) and (*tails, tails*), each of which generates coordination. A backward-induction analysis of the sequential form of Hi-Lo shows that P2 should respond to *high* by *high* and to *low* by *low*. Thus, since P1 prefers the outcome of (*high, high*) to that of (*low, low*), his uniquely optimal choice is *high*; the only permissible path is (*high, high*).[19]

So, if the backward-induction analysis is correct, the sequential forms of Heads and Tails and Hi-Lo do not pose problems for the individually instrumental players of classical game theory. Of Bacharach's three puzzles, only one—the Prisoner's Dilemma—has a sequential variant which need trouble classical theorists. Still, team reasoning is applicable to all three sequential games. For individuals who identify with the relevant group, team reasoning recommends (*cooperate, cooperate*) in the sequential Prisoner's Dilemma, and endorses the backward-induction solutions of the other two games.

But *is* backward-induction analysis valid? Putting the question more precisely: is it true in general that the conclusions of backward-induction analysis are valid implications of the assumptions of classical game theory? It seems that Bacharach thought not. Unfortunately, very little of the reasoning that led him to this conclusion is recorded in his notes. The conclusion itself is stated indirectly in the Scientific Synopsis as part of Bacharach's proposed analysis of 'the person as a team', which was dropped from his later plans for the book. For the present, it is sufficient to say that Bacharach treats the decisions that a person makes over time as if they were made by a sequence of 'transient agents', each representing that person at a separate moment in time. (We say more about this idea in section 7.) As one possible 'architecture' for the reasoning of such agents, Bacharach suggests 'information transfer (memory) with best-reply reasoning'. The idea is that each transient agent acts like an individual player in classical game theory, knowing how all previous agents have acted; this knowledge is supplied by the person's memory of past actions. Bacharach states the following theoretical conclusion, which at the time of writing he intended to prove in the book: 'Perhaps surprisingly, . . . the indeterminacy of game-theoretic rationality with non-tied equilibria is not removed by memory'. This claim is supported by an abbreviated discussion of a sequential-move Hi-Lo game played by five transient agents.[20] The implication seems to be that in sequential Hi-Lo games (or, at least, in games of this kind with a sufficiently large number of players) classical game-theoretic reasoning does not necessarily tell each agent to choose *high*—even though the path on which every agent chooses *high* is the only one deemed permissible by backward-induction reasoning.

But what is wrong with backward-induction reasoning? We do not know how Bacharach would have answered this question, but here is one possible answer, which draws on arguments advanced by Ken Binmore (1987), Philip Pettit and Sugden (1989), and Philip Reny (1992), and which we suggest is consistent with the rest of Bacharach's analysis.[21]

Consider the following four-player sequential version of Hi-Lo. The players, P1, P2, P3 and P4, move in numerical order, each choosing *high* or *low*. Suppose that the path (*high, high, high, high*) gives a payoff of 5 to each player, that (*low, low, low, low*) gives a payoff of 4 to each player, and that, for every other path, the payoff to each player is equal to the number of *high* moves in that path. Given these payoffs, backward-induction reasoning leads to the conclusion that the only permissible path is (*high, high, high, high*).

This reasoning can be represented more explicitly by using the following four premises:

1. Whatever moves P1, P2 and P3 have made, P4 maximises her expected payoff given those moves.

2. Whatever moves P1 and P2 have made, P3 maximises his expected payoff given those moves and given beliefs that are consistent with (1).

3. Whatever move P1 has made, P2 maximises her expected payoff given those moves and given beliefs that are consistent with (1) and (2).

4. P1 maximises his expected payoff, given beliefs that are consistent with (1), (2) and (3).

Using (1), we can deduce that P4 chooses *low* if P1, P2 and P3 have all chosen *low*, and *high* otherwise. Using (2) together with what has already been proved, we can deduce that P3 chooses *low* if P1 and P2 have both chosen *low*, and *high* otherwise. Using (3) together with what has already been proved, we can deduce that P2 chooses *low* if P1 has chosen *low*, and *high* otherwise. Using (4) together with what has already been proved, we can deduce that P1 chooses *high*. Thus the moves chosen by the players trace out the path (*high, high, high, high*).

The question we have to ask is whether all of the premises (1), (2), (3) and (4) can be justified as implications of the classical game-theoretic assumption that it is common knowledge that the players are rational (in the standard, individually instrumental sense of 'rational'). In effect, (1) says that, whatever moves P1, P2 and P3 have made, P4 acts rationally. Let us accept this premise as unproblematic.[22] If we accept this, we should also accept that, whatever moves P1 and P2 have made, P3 acts rationally, given his beliefs: that is one component of (2). Similarly, we should accept that, whatever move P1 has made, P2 acts rationally, given her beliefs: that is a component of (3). And we should accept that P1 acts rationally, given his beliefs: that is a component of (4). But we still need to justify the restrictions that (2), (3) and (4) place on players' beliefs.

For example, consider (2). This tells us that, whatever moves P1 and P2 have made, P3 believes that, whatever P3 himself goes on to do, P4 will act rationally. Here we run into a problem: the clause 'whatever moves P1 and P2 have made' can refer to a counterfactual state of affairs. Suppose it can be proved that, if P1 acts rationally, he chooses *high*. Then, by the assumption that rationality is common knowledge, P1 chooses *high*. So, when it is P3's turn to move, he knows that P1 has chosen *high*. That P3 knows this is a property of the 'actual world'—of how things really are. But premise (2) entails that, had P1 chosen *low*, P3 would still have believed that P4 was rational. The event that P1 chooses *low* belongs to a counterfactual world—a world that can be imagined, but which is not actual. In this counterfactual world, the assumption of common knowledge of rationality does not hold (since P1's choosing *low* can come about only if P1 does not act rationally). So, by virtue of that assumption, we are not entitled to assume that, in that counterfactual world, P3 knows that P4 will act rationally. Presumably there is some counterfactual explanation for P1's irrationality; for all we know, that explanation, whatever it is, might apply to P4 too. To put this point more generally: from the premise that a person knows that two propositions A and B are both true, we are not entitled to infer that, were she to know A to be false, she would still know B to be true.

If this argument is accepted, premise (2) is unjustified. A similar argument applies to (3). What is wrong with these premises is that they assert that, *whether or not previous players had acted rationally*, the player whose turn it is to move would know that every player who moves after her will act rationally. On the strength of the classical assumption of common knowledge of rationality, all we are entitled to say is that *if previous players have acted rationally*, the player whose turn it is to move knows that every player who moves after her will act rationally. However, there is a problem in saying only that. In their original form, (1) to (4) were premises from which we deduced conclusions about the moves that rational players would make. It now seems that, if we use acceptable versions of (2) and (3), backward-induction reasoning may require propositions about the rationality or irrationality of particular moves as the premises from which conclusions about the rationality or irrationality of moves are inferred. This immediately suggests a possible circularity—or, more precisely, that valid backward-induction reasoning may lead only to conclusions about the internal consistency of propositions about rationality, rather than to unconditional conclusions about how it is rational for players to act.

To explore what can legitimately be derived by backward-induction reasoning, we define a *common-knowledge-of-rationality (CKR) path* in a sequential game as a path, every move of which can be shown to be rational by using some mode of reasoning whose validity is common knowledge among the players. Notice that this definition does not specify what modes of reasoning *are* common knowledge. Nevertheless, it allows us to say that, as long as players move along CKR paths, their actions are consistent with the assumption that the players' rationality is common knowledge. Thus we are entitled to assume that common knowledge of rationality is maintained when players move along CKR paths.

We can now restate the premises of backward-induction reasoning, as applied to the four-player Hi-Lo game, as:

1a. Whatever moves P1, P2 and P3 have made, P4 maximises her expected payoff given those moves.

2a. Whatever moves P1 and P2 have made, P3 maximises his expected payoff given those moves and his beliefs; if P1 and P2 have made moves on a CKR path, P3's beliefs are consistent with his knowing (1a).

3a. Whatever move P1 has made, P2 maximises her expected payoff given that move and her beliefs; if P1 has made a move on a CKR path, P2's beliefs are consistent with her knowing (1a) and (2a).

4a. P1 maximises his expected payoff, given beliefs that are consistent with his knowing (1a), (2a) and (3a).

For any path in the game, we can ask whether the proposition that it is a CKR path is consistent with these premises.

It turns out that the proposition that (*low, low, low, low*) is a CKR path *is* consistent with those premises. To prove this result, we make the supposition that (*low, low, low, low*) is a CKR path and then show that, given this supposition, (1a) to (4a) are satisfied along this path.

First, consider P4. Given that the sequence of moves by P1, P2 and P3 has been (*low, low, low*), *low* is P4's payoff-maximising move. So (1a) is satisfied.

Now consider P3. Given that the sequence of moves by P1 and P2 has been (*low, low*), and given that (*low, low, low, low*) is a CKR path,

(2a) requires that P3's beliefs be consistent with his knowing that, whatever move P3 makes, P4's move will be payoff-maximising. So (2a) requires P3 to believe that, were he to choose *low*, P4 would choose *low*. That belief is sufficient to make *low* payoff-maximising for P3. So (2a) is satisfied.

Now consider P2. Given that P1 has chosen *low*, and given that (*low*, *low*, *low*, *low*) is a CKR path, (3a) requires that P2's beliefs be consistent with her knowing that, whatever moves P2 and P3 make, P4's move will be payoff-maximising. So (3a) requires P2 to believe that, were she and P3 both to choose *low*, P4 would choose *low*. It also requires that P2's beliefs be consistent with her knowing that, whatever move she makes, P3's move will be payoff-maximising relative to his (P3's) beliefs and, in addition, that if P2's move is on a CKR path, P3 will know that P4's move will be payoff-maximising. So (3a) requires P2 to believe that, were she to choose *low*, P3 would choose *low*. These beliefs are sufficient to make *low* payoff-maximising for P2. So (3a) is satisfied.

Finally, consider P1. An argument on the same lines as those in the preceding two paragraphs shows that (4a) requires P1 to believe that, were he to choose *low*, P2, P3 and P4 would all choose *low* too. So the question of whether (4a) is satisfied hinges on what would happen were P1 to choose *high*. If this would be followed by P2, P3 and P4 all choosing *high*, (4a) is not satisfied. If it would be followed by any other sequence of moves, (4a) *is* satisfied. But, given only the premises at our disposal, we cannot reject either of these possibilities. The problem is that we are not entitled to assume that P1's choosing *high* is on a CKR path. Thus we cannot assume that, were P1 to choose *high*, P2 would expect P3's and P4's moves to be payoff-maximising. For example, we cannot exclude the possibility that, were P1 to choose *high*, P2 would believe the following: that were she to choose *high*, P3 and P4 would choose *low*, while if she were to choose *low*, P3 and P4 would both choose *high*. If P1 believes that this possibility is the case, it is payoff-maximising for him to choose *low*. This completes the proof: the proposition that (*low*, *low*, *low*, *low*) is a CKR path is consistent with the premises (1a) to (4a).

What does this result tell us? It certainly does not tell us that there is some mode of valid reasoning that recommends each of the moves in the path (*low*, *low*, *low*, *low*) to the relevant player. Indeed, nothing in our analysis tells us that the game *has* a CKR path. To show that, one would need to find a mode of individually instrumental reasoning which yields unconditional conclusions about how the players should act. In its original form—that is, using the premises (1) to (4)—the backward-induction argument was intended to generate such conclusions; but we have argued that those premises are unjustified. Our conclusion is negative: as far as we can see, the classical assumption of common knowledge of rationality does not rule out the path (*low, low, low, low*). It seems that Bacharach is right: there are sequential Hi-Lo games for which classic game-theoretic reasoning does not recommend *high* as uniquely rational for each player.

Some readers may be thinking that, even if our argument is logically correct, it is too contrived to take seriously. We have provided a rationalisation for P1's choosing *low*, but (the reader may ask) would any sensible human player find that rationalisation at all credible? But this objection misses the point. Bacharach's whole analysis of Hi-Lo starts from the recognition that, on any commonsense understanding of rationality, *high* is the uniquely rational choice for each player. So we know from the outset that any argument that denies the unique rationality of *high* will be contrary to common sense. What is at issue is whether such arguments are contrary to the assumptions of game theory.

7. The Person as a Team

Bacharach hoped to apply the theory of team reasoning to problems of *dynamic choice*—that is, problems in which one person makes a sequence of decisions over time. In decision theory, it is conventional to analyse problems of dynamic choice as if, at each time *t* at which the person has to make a decision, that decision is made by a distinct *transient agent*, 'the person at time *t*'. Each transient agent is treated as an independent rational decision-maker. The *person*—the entity that continues through time—is just a collection of transient agents, who may happen to have certain common preferences and beliefs. Just as game theory does not attribute agency to groups, so the theory of dynamic choice does not attribute agency to persons. Bacharach's idea was to represent personal agency by modelling the person as a team of transient agents; each transient agent may identify with the person in essentially the same way that, in the theory of team reasoning, individuals may identify with the group to which they belong. Just as the theory of team reasoning distinguishes between the questions 'What should I do?' and 'What should we do?', so the theory of the person as a team distinguishes between 'What should I-now do?' (where 'I-now' is the currently-acting transient agent) and 'What should I do now?' (where 'I' is the continuing person).

Bacharach intended this analysis to serve two purposes. First, and most obviously, it would be useful in understanding dynamic choice. But in addition, it might help to demystify the concept of team reasoning as applied to groups of individuals. If normal human reasoning about sequences of actions over time could be shown to depend on *intra*personal team reasoning, that would make it much harder for game theory to deny the possibility of *inter*personal team reasoning.

Before pursuing Bacharach's idea in detail, it is worth thinking about its intuitive plausibility compared with that of the conventional analysis of dynamic choice. Suppose you are on the east side of a city street. In order to cross the street, you have to perform two actions in sequence. In *period 1* you have to walk from the east side of the street to the middle. Then, in *period 2*, you have to walk from the middle to the west side. What is involved in your rationally crossing the street?

In the perspective of conventional decision theory, there are two transient agents, 'you in period 1' (P1) and 'you in period 2' (P2). The standard method of analysing this kind of problem is by backward induction (or, as it is sometimes called in the context of dynamic choice, *dynamic programming*). In period 1, your reasoning (as P1) will go something like this: 'I can either stay on the east side or go to the middle. If I go to the middle, the transient agent P2 will then have to choose whether to go on to the west side, or to return to the east side. Since I expect P2 to want to be on the west side, I deduce that he would go on rather than go back. So, since I want P2 to get to the west side, I should go to the middle.' In period 2, you (as P2) note that you are in the middle of the street, and reason: 'I would rather be on the west side than the east, so I should go to the west side'. Reasoning like this may get you across the street, but as an account of actual human reasoning, it does not ring true. What is odd about it is the absence of any intention, plan or sense of agency which extends over the whole episode. In neither period can you think of yourself as performing the action of crossing the street: there is no such 'you' that can perceive itself as a continuing agent.

Now suppose instead that both transient agents identify with the person that is you. Then in period 1, you can reason: 'I (the continuing person) want to be on the west side. The best plan for getting there is for me to go to the middle in period 1, and to continue to the west side in period 2. So that is my plan. To fulfil the first part of that plan, I should go to the middle.' In period 2, you reason: 'To fulfil the second part of my plan to cross the street, I should go to the west side'. In both periods, you can think: 'I am crossing the street'. In this kind of reasoning, the actions that a person takes at different moments are perceived as components of plans which extend over time. This is surely a better description of how human beings normally think about crossing roads.[23]

As the example of crossing the street suggests, dealing successfully with everyday problems of dynamic choice often requires that actions carried out at different times are coordinated with one another. Many such problems have a Hi-Lo structure. Here is an example. Jane buys groceries once a week for four weeks. Each week, she chooses one of two supermarkets at which to shop. The *high* supermarket sells better-quality goods than the *low* one. Each gives a loyalty bonus, for which Jane will qualify if and only if she shops at that supermarket in all four weeks. The bonus is bigger at *low*, but not quite big enough to make up for *low*'s demerits. There are four transient agents, whom we will call Jane_1, Jane_2, Jane_3 and Jane_4. Suppose that all four transient agents have the same preferences over the outcomes of sequences of actions, and that these preferences are represented by a payoff function which gives a payoff of 5 to the sequence (*high, high, high, high*), 4 to (*low, low, low, low*) and, for every other path, a payoff equal to the number of *high* choices. A conventional analysis of this dynamic choice problem would treat it as an interaction between four independent transient agents. If we assume that each transient agent remembers the actions of her predecessors, this interaction is a sequential game. In fact, it is exactly the game we used in section 6 to probe the logic of backward-induction reasoning. All four transient agents agree that the best sequence of actions is (*high, high, high, high*). Conventional backward-induction analysis picks out this sequence as the uniquely optimal solution to the problem. But, as we explained in section 6, Bacharach was not convinced of the general validity of that analysis. If his scepticism is justified, the standard theory of dynamic choice fails to explain some apparently straightforward cases of intrapersonal coordination.

Since the critique of backward-induction reasoning is controversial, and since we do not know whether Bacharach would have endorsed that critique in the form in which we have presented it, it may be more useful to focus on cases in which backward-induction reasoning and team reasoning lead to different conclusions. Specifically, we consider cases in which different transient agents have different preferences over the same sequences of actions. When preferences differ in this way, there are possibilities for strategic interaction between transient agents; much of the literature of dynamic choice theory is concerned with such interactions.

First, we need to fill in some theoretical background. To keep things simple, we use a model in which there is just one person and three periods (1, 2 and 3); the transient agents are P1, P2 and P3. In each period t, the individual's consumption is x_t; an array (x_1, x_2, x_3) is a *consumption stream*. (Each x_t can be interpreted as a vector of quantities of consumption of different goods.) In each period t, the transient agent of that period has preferences over all consumption streams. It is normal to assume that transient agents are motivated by some degree of interagent altruism. (Without such an assumption, it would be difficult to explain acts of saving, by which one transient agent forgoes consumption so as to increase the consumption of later agents.) One simple way of modelling interagent altruism is to postulate, for each period t, an *experienced utility* function v_t which assigns a numerical index to each possible x_t; $v_t(x_t)$ is to be interpreted as a measure of the utility of consuming x_t, as actually experienced in period t. For each period t, we also postulate a *lifetime utility* function u_t which represents preferences over consumption streams, as viewed from that period. These functions are assumed to have the following additive form:

$$u_t = \alpha_{t,1}\, v_1(x_1) + \alpha_{t,2}\, v_2(x_2) + \alpha_{t,3}\, v_3(x_3),$$

$$\text{with } \alpha_{t,r} > 0 \text{ for all } t \text{ and } r.$$

We can now ask whether transient agents have common preferences with respect to (what to them are) present and future consumption. As far as behaviour is concerned, preferences with respect to past consumption are of no significance, since past consumption cannot be changed and (by virtue of the assumption that lifetime utility functions are additive) has no effect on the utility that can be derived from future consumption. In a three-period model, the condition for (relevantly) common preferences is $\alpha_{1,3}/\alpha_{1,2} = \alpha_{2,3}/\alpha_{2,2}$. In words, the relative weights given by P1 to experienced utility in periods 2 and 3 should be the same as the relative weights given by P2 to experienced utility in the same periods.[24] One way of satisfying this condition is to define a *discount* parameter β (with $0 < \beta < 1$) and to set $\alpha_{1,1} = \alpha_{2,2} = \alpha_{3,3} = 1$, $\alpha_{1,2} = \alpha_{2,3} = \beta$, $\alpha_{1,3} = \beta^2$, and $\alpha_{2,1} = \alpha_{3,1} = \alpha_{3,2} = 0$; this is a model of *exponential discounting*.

Although exponential discounting is a common assumption in economic models, there is a lot of evidence that human beings do not discount the future in this mathematically convenient way.[25] The psychological mechanism that leads to deviations from exponential discounting was identified by Hume in the eighteenth century:

> In reflecting on any action, which I am to perform a twelve-month hence, I always resolve to prefer the greater good, whether at that time it will be more contiguous or remote; nor does any difference in that particular make a difference in my present intentions or resolutions. . . . But on my nearer approach, those circumstances, which I at first over-look'd, begin to appear, and have an influence on my conduct and affections. A new inclination to the present good springs up, and makes it difficult for me to adhere inflexibly to my first purpose and resolution. (1739–40/1978, p. 536)

Hume's idea can be expressed in our framework as the hypothesis that, for any period t, the weight given to experienced utility in period t, relative to experienced utility in later periods, is higher when consumption streams are viewed from period t than when they are viewed from earlier periods. To illustrate the implications of this idea as simply as possible, we define a *present good* parameter γ (with $\gamma > 1$), and set $\alpha_{1,1} = \alpha_{2,2} = \alpha_{3,3} = \gamma$ and $\alpha_{1,2} = \alpha_{1,3} = \alpha_{2,1} = \alpha_{2,3} = \alpha_{3,1} = \alpha_{3,2} = 1$. The idea is that, in each period t, the lifetime utility function u_t gives equal weight to the experienced utility of each period except t itself; the experienced utility of period t is given greater weight because of its special 'contiguity'.

In a model of this kind, if transient agents reason as individuals, their actions may combine to produce consumption streams that are unambiguously suboptimal. Here is an example. There are three periods. In period 3, Joe takes an examination. In each of periods 1 and 2, he can choose between revising for the examination (*work*) and relaxing (*rest*). The experienced utility of working in any period is –3, while that of resting is 0. In period 3, experienced utility is 0 if Joe has rested on both previous days, 5 if he has rested on one day and worked on the other, and 10 if he has worked on both. We set $\gamma = 2$. Figure C.4 shows the payoffs in lifetime utility to each of the three transient agents Joe_1, Joe_2 and Joe_3 from each combination of actions by Joe_1 and Joe_2. This is a sequential game in which Joe_1 moves first, followed by Joe_2; Joe_3 does not move.

Notice that the payoffs for Joe_1 and Joe_2 are those of a Prisoner's Dilemma, *work* corresponding with *cooperate* and *rest* with *defect*. Thus the interaction between Joe_1 and Joe_2 is a sequential Prisoner's Dilemma, isomorphic with the game between Hume's farmers (discussed in section 6); the only additional feature is the presence of Joe_3 as a dummy player. From the viewpoint of each of the three transient agents, the sequence of actions (*work, work*) is better than (*rest, rest*). Nevertheless, the backward-induction solution is (*rest, rest*).

As an aside, consider this game from the viewpoint of Joe_1. From Joe_1's viewpoint, the best sequence of actions is (*rest, work*). Reasoning as an individual agent, Joe_1 might naïvely suppose that, since (*rest, work*) is the best sequence from his viewpoint, he ought to choose *rest*. In the literature of dynamic choice, this kind of *myopic* choice is normally treated as irrational, since it involves choosing a plan which a

		Joe_2	
		work	*rest*
Joe_1	*work*	1, 1, 14	–1, 2, 7
	rest	2, –1, 7	0, 0, 0

Figure C.4. Joes's problem (payoffs are shown for Joe_1, Joe_2 and Joe_3)

later transient agent will not continue to follow. Rationality, it is said, requires *sophistication*: each transient agent must treat the predictable choices of future agents as setting feasibility constraints within which he can choose.[26] In Joe's case, Joe_1 has to recognise that the only feasible sequences from which he can choose are (*work, rest*) and (*rest, rest*). Since the latter is better from Joe_1's point of view, he should choose *rest*.

Bacharach's approach allows a different analysis. If Joe_1 and Joe_2 each identify with the person {Joe_1, Joe_2, Joe_3}, each will use team reasoning to determine how he should act. Suppose that the payoff function for the person—the analogue of the payoff function U in the theory of team reasoning—ranks the outcome of (*work, work*) above those of (*work, rest*), (*rest, work*) and (*rest, rest*). (This assumption is analogous with those we have made in previous analyses of Prisoner's Dilemmas; in the present case, Joe_3's payoffs provide additional justification for the assumption.) Then Joe_1 and Joe_2 will each choose *work* as his component of the sequence of actions that is optimal for the person.

8. Resolute Choice

There is some similarity between Bacharach's analysis of the person as a team and Edward McClennen's (1990) analysis of 'resolute choice'. McClennen allows a transient agent to plan a sequence of actions to be taken by himself and by future transient agents and, as an act of will, to resolve to follow that plan. A person is a *resolute chooser* if

> he proceeds, against the background of his decision to adopt a particular plan, to do what the plan calls upon him to do, even though it is true (and he knows it to be true) that were he not committed to choosing in accordance with that plan, he would now be disposed to do something quite distinct from what the plan calls upon him to do. (p. 13)

McClennen argues that resolute choice is psychologically possible for human beings and that 'from a consequentialist perspective itself there is a case to be made for being resolute in a certain kind of situation'—specifically, in situations in which resolute choice furthers the common ends of the transient agents who engage in it (pp. 209–18).

Bacharach saw his team-reasoning theory of dynamic choice as 'in harmony with' McClennen's analysis and as offering 'a new interpretation of resoluteness'.[27] However, Bacharach's analysis differs from McClennen's in not treating the question of whether or not transient agents identify with the person as a matter for rational choice or consequentialist evaluation. For Bacharach, consequentialist evaluation is possible only after the unit of agency has been defined—that is, not until the question 'Consequences for whom?' has been answered. Consequentialism for persons provides a justification for resoluteness, but consequentialism for transient agents does not. In this respect, the difference between Bacharach's and McClennen's analyses of resoluteness parallels a difference between Bacharach's and Gauthier's theories of cooperation in the Prisoner's Dilemma.[28]

It is a central component of resolute choice, as presented by McClennen, that (unless new information becomes available) later transient agents recognise the authority of plans made by earlier agents. Being resolute just *is* recognising that authority (although McClennen's arguments for the rationality and psychological feasibility of resoluteness apply only in cases in which the earlier agents' plans further the common ends of earlier and later agents). This feature of resolute choice is similar to Bacharach's analysis of direction, explained in section 5. If the relationship between transient agents is modelled as a sequential game, resolute choice can be thought of as a form of direction, in which the first transient agent plays the role of director; the plan chosen by that agent can be thought of as a message sent by the director to the other agents. To the extent that each later agent is confident that this plan is in the best interests of the continuing person, that confidence derives from the belief that the first agent identified with the person and that she was sufficiently rational and informed to judge which sequence of actions would best serve the person's objectives.

Bacharach offers an additional account of the role of plans in intrapersonal team reasoning. It is essential to his analysis that whether or not a transient agent identifies with the person is a matter of framing. Thus, just as we can investigate the psychology of group identification for individuals, so we can investigate the psychology of personal identification for transient agents. We can ask what features of a decision problem faced by a transient agent tend to make the personal frame come to mind. Bacharach suggests that the memory of having formed a plan is one such feature.[29] For example, think of Joe_1 and Joe_2. Suppose Joe_1 resolves to follow the plan (*work, work*) and carries out the first component of that plan. Then, when Joe_2 deliberates about whether or not to work, the memory of having previously planned to work may come to mind. This may trigger memories of the reasoning by which he formed that plan. Since that reasoning was based on identification with the person, this may in turn prime concepts connected with personal identification. One implication of this line of thought is that the formation of plans and their committal to memory can be a mechanism by which an earlier transient agent tries to induce a later one to identify with the person. There is a parallel here with Bacharach's analysis of the role of actions which tend to induce others to share one's own sense of group identity: recall his discussion of 'Would you like to dance?'

9. A Last Word

As editors, we have tried to reconstruct as much as we can of Bacharach's theories of framing and team reasoning. We cannot know how far we have been successful in that task. Nor can we know how Bacharach would have developed the ideas he was still working on when he died. Still, our regrets are tempered by the thought that intellectual enquiry is ultimately a cooperative endeavour—a team activity, one might say. Truly valuable contributions to that endeavour are not self-contained theoretical systems that remain the intellectual property of their individual authors. They are living ideas—ideas that take root and grow as part of an evolving body of thought. We are sure that Michael Bacharach's work will live on and we are proud to have helped to record it.

Notes

1. In the language of game theory, this is a *game form*. A game form consists of a set of players, a set of alternative strategies for each player, and, for each profile of strategies that the players might choose, an outcome. In contrast, a *game* is normally defined so that, for each profile of strategies, there is a vector of numerical payoffs, one payoff for each player.

2. Bacharach implicitly endorses this modelling strategy as a norm of game theory when, in a passage from his notes that we quote in section 5, he says 'we must analyse at a level at which there is common knowledge of the payoff matrix'. However, his analysis of framing includes some subtle variations on this strategy. Recall that, if a certain family of predicates does not come to mind for a player, that player reasons as if that family did not exist, even if in fact that family has nonzero 'availability' (that is, nonzero probability of coming to mind). Our concept of 'T-conditional common knowledge' (defined below) is a way of representing these subtleties in the context of circumspect team reasoning.

3. In Bacharach's notes for chapter II, there is a suggestion that some frames have the property that, in order for an agent to have the relevant frame, she must believe that other agents have it too: 'note the curious fact that for some act-descriptions *a* you can't believe you're doing *a* without believing that others are—playing Chess [is] an example. . . . This is an example of a shared or participatory act'. The implication is that if an agent is to conceive of herself as playing Chess, she must believe that her opponent conceives of himself as

playing Chess, and hence that he conceives of her as playing Chess, and so on. Although Bacharach does not say so explicitly, team reasoning in general seems to concern acts that are 'shared' in this sense.

4. This schema is modelled on one which Bacharach uses in his notes for chapter VIII. Bacharach's schema applies to a case in which there is a set of individuals *G* and a 'choice function over profiles' *R*. We take him to be using the standard definition of a 'choice function' as a rule which, for any set of objects of choice—in this case, profiles—picks out a subset, the set of 'choice-worthy' objects. A choice function is slightly more general than a preference ordering, which in turn is slightly more general than a utility or payoff function; but each of these three concepts plays essentially the same theoretical role: it is a way of representing the objectives or rules which underlie actual or recommended choices. Bacharach's schema has two premises: (1) 'It is common knowledge in *G* that *R* is shared in *G*' and (2) 'It is common knowledge in *G* that all in *G* know their components in the best profile'. The conclusion is: 'All choose their components'.

5. Here the distinction between profiles and outcomes is significant. There are a huge number of alternative profiles of actions for players in a football match, but (if draws are resolved by penalty shootouts), there are only two outcomes—win and lose. Because *U* is defined over outcomes, it is undemanding to assume common knowledge of *U*; but a player may know *U* without knowing which profile maximises the expected value of this function. In contrast, if the payoff function had been defined over the profiles themselves, anyone who knew *U* would thereby have known the best profile.

6. We can normalise the payoff function by setting $u_C = 1$ and $u_D = 0$. Then, given that $u_F < 0$, the critical value of ω is $\omega^* = 2u_F/(2u_F - 1)$. The protocol (*cooperate, cooperate*) is optimal if and only if $\omega \geqslant \omega^*$, (*defect, defect*) is optimal if and only if $\omega \leqslant \omega^*$. There is no nonzero value of ω at which (*cooperate, defect*) or (*defect, cooperate*) is optimal.

7. There is some tension between the treatment of the 'we' and 'I' frames in Bacharach's theory of team reasoning and the analysis of frames in his variable frame theory. In variable frame theory, an individual's frame is the set of concepts (strictly, the set of families of concepts) that are available to her; her decision problem is fully determined by the interplay of this frame and the objective features of the world. In the theory of team reasoning, an individual who reasons in the 'we' frame is aware of the 'I' frame too (as one that other players might use), but acknowledges only 'we' reasons. It seems that group identification involves something more than framing in the sense of variable frame theory: the group-identifier does not merely become aware of group concepts, she also becomes committed to the priority of group concepts over individual ones.

8. Rabin's formulation of this hypothesis is not fully compatible with conventional game theory, since it allows each player's utility to depend on his beliefs about other players' beliefs about the first player's choices. (These second-order beliefs are used in defining the first player's beliefs about the second player's intentions.) Rabin's theory uses a nonstandard form of game theory, *psychological game theory* (Geanokoplos, Pearce and Stachetti 1989).

However, as Levine (1998) shows, the main features of Rabin's theory can be reconstructed within conventional game theory.

9. Scientific Synopsis, section 12.5.

10. Sugden (1991) argues for a reinterpretation of constrained maximisation that is more analogous with Bacharach's subsequent treatment of team thinking. The idea is that decision problems can be framed either as choices among actions or as choices among dispositions. Instrumental rationality can be applied within either of these frames, but cannot be used to choose between the frames themselves.

11. See, for example, the papers in Gauthier and Sugden (1993).

12. Bacharach's notes for chapter IX.

13. Scientific Synopisis, section 2.3.

14. In this section, unattributed quotations are from Bacharach's notes for chapter VIII.

15. This example is modelled on the various oracle games Bacharach considers in the notes for chapter VIII, but its specific form is our invention.

16. If instead we define F_2 = {*recommended, nonrecommended*}, we generate another act-description, namely *pick a nonrecommended*. This would not affect the conclusions of the analysis of the current example, in which there is more than one nonrecommended restaurant. However, in cases in which there are only two actions (as in the example of calling runs in cricket), the argument for following the oracle's recommendation breaks down if F_2 includes the predicate *nonrecommended*. Compare the analysis of Heads and Tails in section 4 of the introduction.

17. 'Forward, the Light Brigade!' / Was there a man dismay'd? / Not tho' the soldier knew / Some one had blunder'd: / Theirs not to make reply, / Theirs not to reason why, / Theirs but to do and die, / Into the valley of Death / Rode the six hundred.

18. In the language of game theory, such games are finite extensive-form games with perfect information.

19. These conclusions about Heads and Tails and Hi-Lo are instances of the following result, which applies to any sequential game. Suppose there exists a nonempty set P of paths such that (i) for all paths p, p' in P, the outcomes of p and p' are ranked equally by all players and (ii) for all paths p, p'' where p is in P and p'' is not, the outcome of p is ranked above the outcome of p'' by all players. Then P is the set of permissible paths.

20. These quotations are from section 13.3 of the Scientific Synopsis. In a footnote to this discussion, Bacharach considers a game with transient agents 1, . . . 5, acting in numerical order. Each chooses a or b. The payoff function for all agents is U. Bacharach writes: 'Suppose there are two plans, A and B, each a coordination equilibrium, and A better than B. It seems obvious that agent 1 should follow A, and that, after this, knowing that A has been followed to date supplies a decisive reason for sticking to A. But this need not be so. Let there be five agents, with $A = (a,a,a,a,a)$, $B = (b,b,b,b,b)$. It may well be that $U(a,b,b,b,b) >$ U(a, a,b,b,b); even though 1 followed A, if 2 thinks 3, 4 and 5 will follow B, it is better to switch to B. In such a payoff structure, even though (a,b,b,b,b) is not an equilibrium, it is rationalizable. (1 thinks falsely that 2 thinks the rest will

play *a*, that 2 thinks 3 thinks the rest will . . . , and so that 2 will play *a*; 2 thinks rightly that the rest will play *b*, that 3 thinks the rest will. . . .) What is true is that, after 1 plays *a*, the best *complete sequel* is (*a*,*a*,*a*,*a*). But this is not a compelling reason for 2 unless she is a team reasoner.'

21. For an alternative argument, offered as a defence of backward induction, see Sobel (1993). Sobel argues that the classic game-theoretic assumption of common knowledge of rationality should be interpreted as implying that it is common knowledge, not only that each player is *in fact* rational, but also that each player is 'resiliently' rational. A resiliently rational player acts rationally in any event that is possible within the rules of the game, including those events that cannot be reached by rational play. If there is common knowledge of resilient rationality, the problems discussed in the remainder of this section do not arise.

22. In treating (1) as unproblematic, we are implicitly assuming that players have common knowledge of the true payoff structure of the game, even in counterfactual events that are inconsistent with common knowledge of rationality. If this premise is accepted, the first step of backward-induction reasoning is justified; one implication of this is that backward-induction reasoning is valid for sequential games in which there only two moves. Conversely, if the premise is rejected, even the first step of backward-induction reasoning cannot be justified.

23. On the significance of intentions and plans in practical reason, see Bratman (1987).

24. The general condition is that, for all transient agents *s*, *t*, and for all periods *r* such that $r \geqslant t, s$: $\alpha_{t,r+1}/\alpha_{t,r} = \alpha_{s,r+1}/\alpha_{s,r}$.

25. See, for example, Frederick, Loewenstein and O'Donaghue (2002).

26. This idea has been central to the theory of dynamic choice since the work of Strotz (1955–56).

27. Scientific Synopsis, sections 13.4 and 9.2.

28. For more on this issue, see Sugden (1991).

29. Scientific Synopsis, section 13.4.

References

Aiello, Leslie, and Robin Dunbar. 1993. 'Neocortex Size, Group Size and the Evolution of Language'. *Current Anthropology* 34, 184–93.

Ainslie, George. 1992. *Picoeconomics*. Cambridge: Cambridge University Press.

Anderlini, Luca. 1999. 'Communication, Computability and Common Interest Games'. *Games and Economic Behavior* 27, 1–37.

Arrow, Kenneth. 1982. 'Risk Perception in Psychology and Economics'. *Economic Inquiry* 20, 1–9.

Aumann, Robert J., and Sylvian Sorin. 1989. 'Cooperation and Bounded Recall'. *Games and Economic Behavior* 1, 5–39.

Bacharach, Michael. 1976. *Economics and the Theory of Games*. London: Macmillan.

———. 1987. 'A Theory of Rational Decision in Games'. *Erkenntnis* 27, 17–55.

———. 1991. 'Games with Concept-Sensitive Strategy Spaces'. Paper prepared for International Conference on Game Theory, Florence. Headed 'Draft for comments' and dated April.

———. 1993. 'Variable Universe Games'. In K. Binmore, A. Kirman and P. Tani, eds., *Frontiers of Game Theory*. Cambridge: MIT Press.

———. 1998. 'Preferenze razionali e descrizioni'. In M. Galarotti and G. Gambetta, eds. *Epistemologia et Economia*. Bologna: CLUEB.

———. 1999. 'Interactive Team Reasoning: A Contribution to the Theory of Cooperation'. *Research in Economics* 53, 117–47.

———. 2001. 'Superagency: Beyond an Individualistic Theory of Games'. In Johan van Benthem, ed., *Proceedings of the 8th Conference on Theoretical Aspects of Rationality and Knowledge (TARK-2001)*, San Francisco: Morgan Kaufmann. 333–37.

———. 2003. 'Framing and Cognition: The Bad News and the Good'. In N. Dimitri, M. Basili and I. Gilboa, eds., *Cognitive Processes and Economic Behaviour*. London: Routledge.

Bacharach, Michael, and Michele Bernasconi. 1997. 'The Variable Frame Theory of Focal Points: An Experimental Study'. *Games and Economic Behavior* 19, 1–45.

Bacharach, Michael, and Andrew Colman. 1997. 'Payoff Dominance and the Stackelberg Heuristic'. *Theory and Decision* 43, 1–19

Bacharach, Michael, and Dale O. Stahl. 2000. 'Variable-Frame Level-N Theory'. *Games and Economic Behaviour* 32(2), 220–46.

Bernheim, B. Douglas. 1984. 'Rationalizable Strategic Behaviour'. *Econometrica* 52(4), 1007–28.

Binmore, Ken. 1987. 'Modelling Rational Players: Part I'. *Economics and Philosophy* 3, 9–55.

———. 1992. *Fun and Games: A Text on Game Theory*. Lexington, MA: D. C. Heath.

Binmore, Ken. 1994. *Playing Fair* . Cambridge: MIT Press.
Bjerring, Andrew. 1977. 'The Tracing Procedure and a Theory of Rational Interaction'. In C. A. Hooker, J. Leach and E. McClennen, eds., *Foundations and Applications of Decision Theory*. Dordrecht: Reidel.
Blake, Robert R., and Jane Srygley Mouton. 1986. 'From Theory to Practice in Inter-face Problem Solving'. In S. Worchel and W. Austin, eds., *Psychology of Intergroup Relations*. Chicago: Nelson Hall.
Boehm, Christopher. 1993. 'Egalitarian Society and Reverse Dominance Hierarchy'. *Current Anthropology* 34, 227–54.
Boesch, Christophe, and Hedwige Boesch. 1989. 'Hunting Behavior of Wild Chimpanzees in the Tai National Park'. *American Journal of Physical Anthropology* 78, 547–73.
Bolton, Gary, and Axel Ockenfels. 2000. 'ERC—a Theory of Equity, Reciprocity and Competition'. *American Economic Review* 90, 166–93.
Bornstein, Gary, Uri Gneezy and Rosemary Nagel. 2002. 'The Effect of Intergroup Competition on Group Coordination: An Experimental Study'. *Games and Economic Behavior* 41(1), 1–25.
Boyd, Robert, and Peter Richerson. 1985. *Culture and the Evolutionary Process*. Chicago: University of Chicago Press.
Brain, Charles K. 1981. *The Hunters or the Hunted?* Chicago: University of Chicago Press.
Bratman, Michael E. 1987. *Intentions, Plans and Practical Reason*. Cambridge, MA: Harvard University Press.
———. 1993. 'Shared Intention'. *Ethics* 104, 97–113.
Brewer, Marilynn B. 1991. 'The Social Self: On Being the Same and Different at the Same Time'. *Personality and Social Psychology Bulletin* 17, 475–82.
Brewer, Marilynn B., and Wendi Gardner. 1996. 'Who Is This "We"? Levels of Collective Identity and Self Representations'. *Journal of Personality and Social Psychology* 71(1), 83–93.
Brewer, Marilynn B., and Roderick M. Kramer. 1986. 'Choice Behavior in Social Dilemmas: Effects of Social Identity, Group Size, and Decision Framing'. *Journal of Personality and Social Psychology* 50(3), 543–49.
Brewer, Marilynn B., and Norman Miller. 1996. *Intergroup Relationships*. Buckingham, U.K.: Open University Press.
Broome, John. 1991. *Weighing Goods*. Oxford: Blackwell.
Bruner, Jerome. 1957. 'On Perceptual Readiness'. *Psychological Review* 64, 123–51.
Camerer, Colin. 1995. 'Individual Decision Making'. In John Kagel and Alvin Roth, eds., *The Handbook of Experimental Economics*. Princeton, NJ: Princeton University Press.
Campbell, Donald T. 1958. 'Common Fate, Similarity and Other Indices of the Status of Aggregates of Persons as Social Entities'. *Behavioral Science* 3, 14–25.
———. 1994. 'How Individual and Face-to-Face Group Selection Undermine Firm Selection in Organizational Evolution'. In Joel A. C. Baum and Jitendra V. Singh, eds., *Evolutionary Dynamics of Organizations*, 23–38. Oxford: Oxford University Press.

Caporael, Linnda R. 1995. 'Sociality: Coordinating Bodies, Minds and Groups'. *Psycoloquy* 6. princeton.edu Directory:pub/harnad/Psycoloquy/1995.volume.6
Casajus, André. 2001. *Focal Points in Framed Games: Breaking the Symmetry*. Berlin: Springer-Verlag.
Collard, David. 1978. *Altruism and Economy*. Oxford: Martin Robertson.
Colman, Andrew, and Jonathan Stirk. 1998. 'Stackelberg Reasoning in Mixed-Motive Games: An Experimental Investigation'. *Journal of Economic Psychology* 19, 279–93.
Cookson, Richard. 2000. 'Framing Effects in Public Goods Games'. *Experimental Economics* 3, 55–79.
Cosmides, Leda, and John Tooby. 1989. 'Evolutionary Psychology and the Generation of Culture, Part II, Case Study: A Computational Theory of Social Exchange'. *Ethology and Sociobiology* 10, 51–79.
Crawford, Vince, and Hans Haller. 1990. 'Learning How to Cooperate: Optimal Play in Repeated Coordination Games'. *Econometrica* 58, 571–95.
Dawes, Robyn M., John Orbell and Alphons van de Kragt. 1990. 'The Limits of Multilateral Promising'. *Ethics* 100, 616–27.
Dawes, Robyn M., Alphons van de Kragt and John Orbell. 1988. 'Not Me or Thee but We: The Importance of Group Identity in Eliciting Cooperation in Dilemma Situations: Experimental Manipulations'. *Acta Psychologica* 68, 83–97.
De Cremer, David, and Mark Van Vugt. 1999. 'Social Identification Effects in Social Dilemmas: A Transformation of Motives'. *European Journal of Social Psychology* 29, 871–93.
Dennett, Daniel C. 1987. 'Intentional Systems in Cognitive Ethology: The "Panglossian Paradigm" Defended'. In Dennett, *The Intentional Stance*, 237–88. Cambridge: MIT Press.
Diener, Edward. 1977. 'Deindividuation: Causes and Consequences'. *Social Behavior and Personality* 5(1), 143–55.
Dion, Kenneth L. 1973. 'Cohesiveness as a Determinant of Ingroup-Outgroup Bias'. *Journal of Personality and Social Psychology* 28(2), 163–71.
Doise, Willem, György Csepeli, H. D. Dann, C. Gouge, K. S. Larsen and Alistair Ostell. 1972. 'An Experimental Investigation into the Formation of Intergroup Representations'. *European Journal of Social Psychology* (2), 202–4.
Dunbar, R. I. M. 1993. 'Coevolution of Neocortical Size, Group Size and Language in Humans'. *Behavioral and Brain Sciences*, 16(4), 681–735.
Durkheim, Emile. 1893. *La division du travail social*. Paris: Presses Universitaires de France.
Elster, Jon. 1979. *Ulysses and the Sirens: Studies in Rationality and Irrationality*. Cambridge: Cambridge University Press.
———. 1989. *The Cement of Society*. Cambridge: Cambridge University Press.
Fagin, Ronald, Joseph Y. Halpern, Yoram Moses and Moshe Y. Vardi. 1995. *Reasoning about Knowledge*. Cambridge: MIT Press.
Falk, Armin, and Urs Fischbacher. 1999. 'A Theory of Reciprocity'. Working paper 6. Institute for Empirical Research in Economics, University of Zurich.

Farrell, Joseph. 1988. 'Communication, Coordination and Nash Equilibrium'. *Economics Letters*, 27, 209–14.
Fehr, Ernst, and Klaus Schmidt. 1999. 'A Theory of Fairness, Competition and Cooperation'. *Quarterly Journal of Economics* 114, 817–68.
Frederick, Shane, George Loewenstein and Ted O'Donaghue. 2002. 'Time Discounting and Time Preference: A Critical Review'. *Journal of Economic Literature* 40, 351–401.
Friedman, Daniel. 1991. 'Evolutionary Games in Economics'. *Econometrica* 59, 637–66
Gauthier, David. 1975. 'Coordination'. *Dialogue* 14, 195–221.
———. 1986. *Morals by Agreement*. Oxford: Clarendon Press.
Gauthier, David, and Robert Sugden, eds. 1993. *Rationality, Justice and the Social Contract: Themes from 'Morals by Agreement'*. London: Harvester Wheatsheaf.
Geanakoplos, John, David Pearce and Ennio Stachetti. 1989. 'Psychological Games and Sequential Rationality'. *Games and Economic Behavior* 1, 60–79.
Gilbert, Margaret. 1989. *On Social Facts*. London: Routledge.
———. 1997. 'What Is It for Us to Intend?' in H. Holmstrom-Nintikka and R. Tuomela, eds. *The Philosophy and Logic of Social Action*, vol. 2. Dordrecht: Kluwer.
———. 1998. 'In Search of Sociality'. *Philosophical Explorations* 1(3), 233–41.
Gintis, Herbert. 2000. *Game Theory Evolving*. Princeton, NJ: Princeton University Press.
Grice, Herbert. P. 1989. *Studies in the Way of Words*. Cambridge, Mass.: Harvard University Press.
Gurin, Patricia, and Hazel Markus. 1988. 'Group Identity: The Psychological Mechanisms of Durable Salience'. *Revue Internationale de Psychologie Sociale* 1, 257–74.
Hamilton, William D. 1963. 'The Evolution of Altruistic Behavior'. *American Naturalist* 97, 354–66.
———. 1964. 'The Genetical Evolution of Social Behaviour, II', *Journal of Theoretical Biology* 7, 17–52.
———. 1975. 'Friends, Romans, Groups . . . : Innate Social Aptitudes of Man: An Approach from Evolutionary Genetics'. In *Narrow Roads of Gene Land: The Collected Papers of W. D. Hamilton*. Oxford: W. H. Freeman.
Hardin, Russell. 1995. *One for All: The Logic of Conflict*. Princeton, NJ: Princeton University Press.
Harsanyi, John. 1975. 'The Tracing Procedure'. *International Journal of Game Theory* 4, 61–94.
———. 1980. 'Rule Utilitarianism, Rights, Obligations and the Theory of Rational Behaviour'. *Theory and Decision* 12, 115–33.
Harsanyi, John, and Reinhard Selten. 1988. *A General Theory of Equilibrium Selection in Games*. Cambridge: MIT Press.
Hatfield, Elaine, John T. Cacioppo and Richard L. Rapson. 1994. *Emotional Contagion*. Cambridge: Cambridge University Press.
Hodgson, David. 1967. *Consequences of Utilitarianism*. Oxford: Clarendon Press.
Hollis, Martin. 1998. *Trust within Reason*. Cambridge: Cambridge University Press.

Holmstrom, Bengt. 1982. 'Moral Hazard in Teams'. *Bell Journal of Economics* 13, 324–40.

Huczynski, Andrezej, and David Buchanan. 1985. *Organizational Behaviour*. London: Prentice Hall.

Hume, David. 1739–40/1978. *A Treatise of Human Nature*. Oxford: Clarendon Press.

Hurley, Susan. 1989. *Natural Reasons*. New York: Oxford University Press.

Jacobsen, Hans Jorgen. 1996. 'On the Foundations of Nash Equilibrium'. *Economics and Philosophy* 12, 67–88.

Janssen, Maarten C. W. 2001a. Towards a Justification of the Principle of Coordination'. *Economics and Philosophy* 17(2), 221–34.

———. 2001b. 'Rationalizing Focal Points'. *Theory and Decision* 50, 119–48.

Jones, John E. 1973. 'Model of Group Development'. In *The 1973 Annual Handbook for Group Facilitators*. San Francisco: Pfeiffer/Jossey-Bass.

Kahneman, Daniel, and Amos Tversky. 1979. 'Prospect Theory: An Analysis of Decision under Risk'. *Econometrica* 47, 263–91.

Kitcher, Philip. 1993. 'The Evolution of Human Altruism'. *Journal of Philosophy*, 90(10), 497–516.

Kramer, Roderick M., and Marilynn B. Brewer. 1984. 'Effects of Group Identity on Resource Use in a Simulated Commons Dilemma'. *Journal of Personality and Social Psychology*, 46(5), 1044–57.

Laffont, Jean-Jacques. 1975. 'Macroeconomic Constraints, Economic Efficiency and Ethics: An Introduction to Kantian Economics'. *Economica* 42, 430–37.

Levine, David. 1998. 'Modelling Altruism and Spitefulness in Experiments'. *Review of Economic Dynamics* 1, 593–622.

———. 1979. 'Prisoner's Dilemma Is a Newcomb Problem'. *Philosophy and Public Affairs* 8, 235–40.

Lewis, David K. 1969. *Convention: A Philosophical Study*. Cambridge, MA: Harvard University Press.

Lipsey, Richard G., and Lancaster Kelvin. 1956. 'The General Theory of the Second-Best'. *Review of Economic Studies* 24, 11–32.

Lott, Albert J., and Bernice E. Lott. 1965. 'Group Cohesiveness as Interpersonal Attraction'. *Psychological Bulletin* 64, 259–309.

Marschak, Jacob, and Radner, Roy. 1972. *Economic Theory of Teams*. New Haven, CT: Yale University Press.

Maynard Smith, John. 1982. *Evolution and the Theory of Games*. Cambridge: Cambridge University Press.

Mayo, Elton. 1933. *The Human Problems of an Industrial Civilization*. New York: Macmillan.

McClennen, Edward. 1990. *Rationality and Dynamic Choice: Foundational Explorations*. Cambridge: Cambridge University Press.

Mehta, Judith, Chris Starmer and Robert Sugden. 1994. 'The Nature of Salience: An Experimental Investigation of Pure Coordination Games'. *American Economic Review* 84, 658–73.

Mithen, Steven. 1996. *The Prehistory of the Mind*. London: Thames and Hudson.

Myerson, Roger. 1991. *Game Theory: Analysis of Conflict*. Cambridge, MA: Harvard University Press.

Nash, John. 1950. 'The Bargaining Problem'. *Econometrica* 18, 155–62.
———. 1953. 'Two-Person Cooperative Games'. *Econometrica* 21, 128–40.
Oakes, Penelope, Alexander Haslam and John Turner. 1994. *Stereotyping and Social Reality*. Oxford: Blackwell.
Offerman, Theo, Joep Sonnemans and Arthur Schram. 1996. 'Value Orientations, Expectations and Voluntary Contributions in Public Goods'. *Economic Journal* 106, 817–45.
Pearce, David G. 1984. 'Rationalizable Strategic Behaviour and the Problem of Perfection'. *Econometrica* 52(4), 1029–50.
Perdue, Charles W., John F. Dovidio, Michael B. Gurtman and Richard B. Tyler. 1990. 'Us and Them: Social Categorization and the Process of Intergroup Bias'. *Journal of Personality and Social Psychology* 59(3): 475–86.
Pettit, Philip, and Robert Sugden. 1989. 'The backward induction paradox'. *Journal of Philosophy* 86: 169–82.
Prentice, Deborah A., and Dale T. Miller. 1992. 'The Psychology of Ingroup Attachment'. Paper presented at conference on The Self and the Collective, Princeton University.
Price, George R. 1970. 'Selection and Covariance'. *Nature* 277, 520–21.
Rabbie, Jacob, and Murray Horwitz. 1969. 'Arousal of Ingroup-Outgroup Bias by a Chance Win or Loss'. *Journal of Personality and Social Psychology* 13(3), 269–77.
Rabbie, J. M., J. C. Schot and L. Visser. 1989. 'Social Identity Theory: A Conceptual and Empirical Critique from the Perspective of a Behavioural Interaction Model'. *Advances in Group Processes* 11, 139–74.
Rabin, Matthew. 1993. 'Incorporating Fairness into Game Theory and Economics'. *American Economic Review* 83, 1281–1302.
Rasmusen, Eric. 1987. 'Moral Hazard in Risk-Averse Teams'. *RAND Journal of Economics* 18(3), 428–35.
Regan, Donald. 1980. *Utilitarianism and Co-operation*. Oxford: Oxford University Press.
Reny, Philip. 1992. 'Backward Induction, Normal Form Perfection and Explicable Equilibria'. *Econometrica* 60, 627–49.
Roethlisberger, Fritz, and William Dickson. 1939. *Management and the Worker*. Cambridge: Cambridge University Press.
Rosch, Eleanor. 1978. 'Principles of Categorization'. In Eleanor Rosch and Barbara Lloyd, eds., *Cognition and Categorization*. Hillsdale, NJ: Erlbaum.
Ross, Marc Howard. 1986. 'A Cross-Cultural Theory of Political Conflict and Violence'. *Political Psychology* 7, 427–69.
Rousseau, Jean-Jacques. 1762/1988. 'On Social Contract'. In Alan Ritter and Julia Conaway Bondanella, eds., *Rousseau's Political Writings*. New York: Norton.
Sally, David. 1995. 'Conversation and Cooperation in Social Dilemmas: A Meta-analysis of Experiments from 1958 to 1992'. *Rationality and Society* 7, 58–92.
Schelling, Thomas. 1960. *The Strategy of Conflict*. Cambridge, MA: Harvard University Press.
Schick, Frederic. 1997. *Making Choices: A Recasting of Decision Theory*. Cambridge: Cambridge University Press.

Searle, John. 1990. `Collective Intentions and Actions'. In P. Cohen, J. Morgan, M. E. Pollack, eds., *Intentions in Communication*. Cambridge: MIT Press.
———. 1995. *The Construction of Social Reality*. London: Allen Lane.
Selten, Reinhard. 1975. 'Reexamination of the Perfectness Concept for Equilibrium Points in Extensive Form Games'. *International Journal of Game Theory* 4, 25–55.
Sethi, Rajiv, and E. Somanathan. 1999. 'Preference Evolution and Reciprocity'. Working paper 99-06, Barnard College, New York.
Sherif, Muzafer, O. J. Harvey, B. Jack White, William R. Hood and Carolyn W. Sherif. 1961. *Intergroup Conflict and Cooperation: The Robbers Cave Experiment*. Norman: University of Oklahoma Book Exchange.
Simmel, Georg. 1910. 'How Is Society Possible?' *American Journal of Sociology* 16(3), 382–91.
Simon, Herbert A. 1979. 'From Substantive to Procedural Rationality'. In F. Hahn and M. Hollis, eds., *Philosophy and Economic Theory*. Oxford: Oxford University Press.
Smith, Eliot R., and Susan Henry. 1996. 'An In-group Becomes Part of the Self: Response Time Evidence'. *Personality and Social Psychology Bulletin* 22, 635–42.
Sobel, J. Howard. 1993. 'Backward Induction Arguments in Finitely Iterated Prisoners, Dilemmas: A Paradox Regained'. *Philosophy of Science* 60: 114–33.
Sober, Elliott, and David Sloan Wilson. 1998. *Unto Others: The Evolution and Psychology of Unselfish Behavior*. Cambridge, MA: Harvard University Press.
Stahl, Dale O., and Wilson, Paul W. 1994. 'Experimental Evidence on Players' Models of Other Players'. *Journal of Economic Behavior and Organization* 25, 309–27.
———. 1995. 'On Players' Models of Other Players—A New Theory and Experimental Evidence'. *Games and Economic Behaviour* 10, 218–54.
Strotz, Robert H. 1955–56. 'Myopia and Inconsistency in Dynamic Utility Maximization'. *Review of Economic Studies* 23, 165–80.
Sugden, Robert. 1984. 'Reciprocity: The Supply of Public Goods through Voluntary Contributions'. *Economic Journal* 94, 772–87.
———. 1991. 'Rational Choice: A Survey of Contributions from Economics and Philosophy'. *Economic Journal* 101, 751–85.
———. 1993. 'Thinking as a Team: Towards an Explanation of Nonselfish Behavior'. *Social Philosophy and Policy* 10, 69–89.
———. 1995. 'A Theory of Focal Points'. *Economic Journal* 105, 533–50.
———. 2000. 'Team Preferences'. *Economics and Philosophy* 16, 175–204.
———. 2003. 'The Logic of Team Reasoning' *Philosophical Explorations* 6, 165–81.
Tajfel, Henri. 1969. 'Cognitive Aspects of Prejudice'. *Journal of Social Issues* 25, 79–97.
———. 1970. 'Experiments in Intergroup Discrimination'. *Scientific American* 223, 96–102.
Tajfel, Henri, and J. P. Forgas. 1981. 'Social Categorisation: Cognitions, Values and Groups'. In J. P. Forgas, ed., *Social Cognition*. London: Academic Press.

Taylor, Frederick W. 1911/1967. *The Principles of Scientific Management*. New York: W. W. Norton.

Trivers, Robert. L. 1971. 'The Evolution of Reciprocal Altruism'. *Quarterly Review of Biology* 46, 35–57.

Tuomela, Raimo. 1995. *The Importance of Us*. Stanford, CA: Stanford University Press.

Turner, John C., Michael Hogg, Penelope J. Oakes, Stephen Reicher and Margaret Wetherell. 1987. *Rediscovering the Social Group: A Self-Categorization Theory*. Oxford: Blackwell.

Tversky, Amos. 1977. 'Features of Similarity'. *Psychological Review* 84(1), 327–52.

Tversky, Amos, and Daniel Kahneman. 1986. 'Rational Choice and the Framing of Decisions'. *Journal of Business* 59, S251–78.

Van de Kragt, Alphons, Robyn M. Dawes, John Orbell, S. R. Braver, and L. A. Wilson II. 1986. 'Doing well and doing good as ways of resolving social dilemmas'. In H. A. M. Wilke, D. M. Messick and C. G. Rutte, eds., *Experimental social dilemmas*, pp. 177–203. Frankfurt am Main: Verlag Peter Lang.

Van Huyck, John B., Raymond C. Battalio and Richard O. Beil. 1990. 'Tacit Coordination Games, Strategic Uncertainty and Coordination Failure'. *American Economic Review* 80(1), 234–48.

Van Winden, Frans. 1983. *On the Interaction between State and Private Sector: A Study in Political Economics*. Amsterdam: North-Holland.

Wertheimer, Max. 1923. 'Untersuchungen zur Lehre von der Gestalt'. *Psychologische Forschung* 4, 301–50.

Wilson, David Sloan. 1997. 'Incorporating Group Selection into the Adaptationist Program: A Case Study Involving Human Decision Making'. In Douglas Kendrick and Jeffrey Simpson, eds., *Evolutionary Social Psychology*. Mahwah, NJ: Erlbaum Press.

———. 1998. 'Hunting, Sharing and Multi-Level Selection: The Tolerated Theft Model Revisited'. *Current Anthropology* 39, 73–97.

Wilson, W., and M. Katayani. 1968. 'Intergroup Attitudes and Strategies in Games between Opponents of the Same or of a Different Race'. *Journal of Personality and Social Psychology* 9, 24–30.

Wright, Sewall. 1945. 'Tempo and Mode in Evolution: A Critical Review', *Ecology* 26, 415–19.

Zermelo, Ernst. 1912. 'Über eine Anwendung der Memgenlehre auf die Theorie des Schachspiels'. *Proceedings of the Fifth International Conference of Mathematicians, Cambridge* 2, 501–10.

Zizzo, Daniel, and Jonathan Tan. 2002. 'Game Harmony as a Predictor of Cooperation in 2 × 2 Games'. Oxford University Department of Economics Working Paper Series, Ref: 151.

Index